D1557480

# The Golden Cell

# The Golden Cell

Karen van Kampen

## GENE THERAPY, STEM CELLS, AND THE QUEST FOR THE NEXT GREAT MEDICAL BREAKTHROUGH

HarperCollinsPublishersLtd

*The Golden Cell*


Published by HarperCollins Publishers Ltd



First edition

HarperCollins books may be purchased for educational, business, or sales promotional use through our Special Markets Department.

HarperCollins Publishers Ltd
2 Bloor Street East, 20th Floor
Toronto, Ontario, Canada
M4W 1A8

*www.harpercollins.ca*

Library and Archives Canada Cataloguing in Publication

van Kampen, Karen, 1962–
The golden cell : gene therapy, stem cells, and the quest for the next great medical breakthrough / Karen van Kampen.

Includes index.
ISBN-13: 978-0-00-200794-8
ISBN-10: 0-00-200794-0

1. Genetic engineering. 2. Genetic engineering industry. I. Title.

RB155.8.V35 2005 660.6 C2005-901679-5

HC 9 8 7 6 5 4 3 2 1

Printed and bound in the United States
Set in Bembo

FOR MY PARENTS,

DR. MURRAY AND LINDA MOFFAT

# Contents

Preface xi
Introduction 1

**1**
**Taxicabs and Roadblocks**
Navigating Genes through the Highways of the Body
3

**2**
**The Gene Hunters**
On the Trail of Deadly Disease Genes
31

**3**
**A New Code to Crack**
Population Genetics and the Genome's Instruction Manual
59

**4**
**Body Language**
Interpreting the Messages of Stem Cells
86

5

**The Body Builders**

Tissue Engineers Reconstruct Broken Bones and Ailing Parts

116

6

**Heal Thyself**

When Religion, Ethics, and Science Meet

143

7

**Golden Cells and Flying Pigs**

The Business of Biotechnology

166

8

**Owning Ourselves (and Each Other)**

The Patenting of Life

196

9

**The Gene Thieves**

Biopiracy in the Developing World

224

Notes 246

Glossary 271

Bibliography 279

Index 285

# Preface

I was on a steep learning curve during the research and writing of this book, which would not have been possible without the time, energy, and attention of many individuals. For their scientific expertise, I would like to thank Dr. Michael Kaplitt, Dr. Martin Kaplitt, Lap-Chee Tsui, Molly Shoichet, John Davies, Michael Rudnicki and his team (namely Pearl Campbell), Jane Aubin and her team (specifically Shulin Zhang), James Till, Dr. Ernest McCulloch, Dr. Ron Worton, Jeffrey Karp, Kathryn Moore, Michael Sefton, Dr. Thomas Okarma, Dr. Roderick McInnes, Dr. Irving Weissman, Michael West, Ananda Mohan Chakrabarty, Richard Rozmahel, Dominique Stoppa-Lyonnet, Dr. Edward Diethrich, Dr. Nabil Dib, and Dr. Andres Metspalu. For their invaluable perspectives on regenerative medicine, I am grateful to Timothy Caulfield, Bartha Maria Knoppers, Halla Thorsteinsdottir, Dr. John Dossetor, Francoise Baylis, and Alan Milstein. For their insights into the book's subjects, I thank Brian Shoichet, Max Cooper, and Miller Quarles. For their personal experiences, I thank Paul Gelsinger and Nathan Klein. For his understanding of the business of biotechnology, I thank Bob Mark. For their navigation of resources, I thank the staff of the University of Toronto's Gerstein Science Information Centre and Bora Laskin Law Library, the Toronto Reference Library, and the Princess Margaret Hospital library. For his deft pen and incisive input, I thank my editor, Jim Gifford. I am indebted to my agents, Don Sedgwick and Shaun Bradley, whose confidence and enthusiasm made this project possible. I thank my brother, Grant Moffat, for his legal expertise and long, inspiring chats. I am grateful to my parents, Dr. Murray and Linda Moffat, for their encouragement, inspiration, and Muskoka writer's retreat. I thank my husband, Dimitri van Kampen, for his meticulous eye for detail and unwavering support.

# Introduction

In June 2002, I was sent by the *National Post* to cover ChaRM, the Challenges in Regenerative Medicine conference. Dozens of scientists and physicians had taken over a ballroom in Toronto's Fairmont Royal York Hotel with their exhibits. It was like a science fair for adults. Colourful sheets of cardboard displayed successful experiments. But rather than a grade-school how-to on coaxing a paper-mâché volcano to erupt, I found detailed instructions for building three-dimensional cellular bridges to repair damaged spines, and a step-by-step guide to create chemical channels using lasers to control nerve regeneration.

There were diagrams explaining how to metamorphose embryonic stem cells into other types of cells. These stem cells are content to remain as they are born, as blank slates free of motivation or foresight. Embryonic stem cells are magical in that they can be coaxed into becoming virtually all of the body's cell types, which could one day be used to strengthen weakened hearts or ailing Parkinson's and Alzheimer's brains.

As I navigated through the maze of exhibits, examining diagrams of hearts, brains, and spines, I realized that the body was a different landscape than I had envisioned. I have always been mystified by its complexities. How do our hearts, organs the size of a fist, manage to beat 100,000 times a day, pumping 10,000 litres of blood through our systems? How do our lungs know to exchange carbon dioxide for oxygen? How do our brains process everything that happens around us, guiding how we feel, what we dream, and what we remember?

The body has always been a familiar as well as an uncharted landscape.

It is as if I am flying over the city in which I live. I can pinpoint landmarks, but everything in between is unrecognizable. I see the body as an aerial map of 100 trillion cells, a map that becomes dizzying when I try to take it in all at once. But if I focus on one section of land, zooming in to one city block, then one street, and finally one house, the picture moves into focus. Then, rather than scanning 100 trillion cells, I am looking at one single cell.

Life, it seems, can be simplified to the intricacies of the cell. Our genetic blueprint, our physical sketch, even our biological destinies are hiding within them. Each cell houses variations of genes that determine our hair and eye colour, the diseases we are born with, and the conditions that we may one day develop.

Regenerative medicine uses what is hiding within all of us—within our own cells—to fight the diseases that may some day threaten our very lives. It replaces faulty genes and corrects renegade proteins. It uses the stem cells floating within our blood, bone, and muscle to mend broken limbs and repair weakened organs. For years, scientists studied the inner workings of the cell, stalking genes and proteins to record their every move. They learned what happens biologically within each of us during a lifetime, and specifically when certain diseases strike. *The Golden Cell* shows scientists posing a fundamental question: What makes a cell tick? Then there are the questions that follow. What are the mechanisms that convince a cell to grow, divide, or change? How does a cancerous cell come into being, multiplying by the millions and wreaking havoc on the body? How does the environment convince genes to mutate?

Like most people, I have seen how a genetic disease can crumple a healthy person, how a serious injury or accident can drain the life out of someone. Imagine if broken spines could be mended, tired organs restored, and addled minds awakened. As Molly Shoichet, co-chair of ChaRM, explained, these are the promises of regenerative medicine. As you will read, the next frontier in medicine can release the regenerative powers of the cell, unlocking new life within us all.

# Chapter One

## Taxicabs and Roadblocks

### Navigating Genes through the Highways of the Body

For five excruciating hours he lay here, pinned to an operating table at New York Weill Cornell Medical Center. He willed every sensation, every impulse, to leave his body. The pins and needles crawling up his legs; the suppressed cough scratching his sandpapery throat. Everything he felt had to be ignored. Even the wild beating of his heart. For five hours, his mind had to forget his body. His thoughts had to shut down, while he had to remain awake. If he surrendered to sleep, the operation would also be surrendered. The months of hope and anticipation would have been for naught.

With the turn of a few screws he had been clamped in place, anchored by bits of steel. He had become as solid and immobile as the operating table. It was as if he had been poured into a mould and set in this fixed position. His concentration, though, was being taunted. There was the slow *whoo-sh* forward and quick *s-nap* back of the swinging door; the monotone *be-eep, be-eep, be-eep* of machines encircling him; and the clink of scissors and scalpels and other disinfected tools as they were lined up on a tray, ready to be used, *gasp*, on him. Just stare at the blank walls, or the blank ceiling, and disappear into this nothingness. The room was a clean slate, and suddenly, so was his life.

It was a muggy August day in 2003, but the operating room was impervious to weather or time. The scene, if caught on film, would reproduce as a blur of movement and colour, streaks of green scrubs staining a white background. It was a maze of bodies darting from one place to another, each following a prescribed course, all in their own world. Physicians, nurses, and technicians, gowned and gloved, popped in and out of view as he lay on his back, his spine kissing the cold stainless steel table. He too was concentrating on what he had to do. He could not be put under general anesthetic because this would lull his brain to sleep, making the operation impossible to perform. It was his job to stay awake for the entire surgery, and not move. No shifting. No twitching. He couldn't even think about coughing or yawning or giving a good stretch of those achingly sedentary legs. One sneeze rattling his body could prove fatal.

But what was at stake was far more complicated. Sure, he knew the surgical risks. They applied to all brain surgery. He could have a bleed. The brain doesn't like being poked and prodded, and if his surgeon guided the needle even a millimetre off target, he could prick one of those wispy floating veins and release a waterfall of blood. The brain is like a powder keg. Pressure from a bleed pushes against its outer walls, crushing anything in its path, and possibly ending in a fatal explosion. He could also get an infection, or his body might reject the therapy. But he also knew what would happen if he hadn't signed up: nothing.

Then there were the possibilities. It wasn't just the dream of loosening the grip of Parkinson's on his withering 55-year-old body. It was the chance of changing the course of medicine; propelling the field of gene therapy; perfecting a treatment for neurological diseases. That was what had compelled him to volunteer for this experimental surgery. Here he was, documentary filmmaker and Parkinson's sufferer Nathan Klein, a man who would help make medical history.

Klein was diagnosed with Parkinson's over a decade ago. No one in his family has the disease. Of course he knew of Parkinson's before his diagnosis, but only in the abstract. He lacked any foresight of what was to come. "It was shocking, but I said, I'll deal with it," says Klein, who did so for many years, alleviating symptoms with handfuls of pills. But over time, the disease began to invade every aspect of his life. Now, daily routines are maddening. Shaving is torture, as is drinking a cup of coffee. And playing ball with his son is close to impossible. "I feel I'm depriving my kids of their father, and it's a little upsetting," says Klein. "But that's the nature of the game."[1]

Born in Israel, Klein moved with his family to New York before he was five. He grew up in the Bronx, competing as a gymnast in high school. It's tough looking back, remembering how agile he once was. Lately he's gone back to the gym to coax his body out of hibernation, but a membership costs money, which these days he has little of.

Klein now lives in Port Washington, Long Island, with his wife, Claire, and their teenage twins, Jacklyn and Eric. They used to own a house, but he needed the money and had to sell. "You've got to support a family. You've got a disease to deal with," he says. "You've got a lot of things to deal with." Klein works as a freelance television producer. It's fortunate that he is in business for himself. He knows of many people who had to leave their company as the disease progressed. But Parkinson's has softened his voice, making it difficult to command a newsroom with a whisper.

Then there are the side effects. His body has rejected a lot of the Parkinson's medication. He has tried so many different prescriptions—some that bloat him beyond his normal frame, others that strip him of kilos and, as he says, make him look like the "night of the living dead." For years, Klein told his neurologist that he would be a guinea pig for experimental treatments. Before this clinical trial, he participated in several studies. Nothing much changed. Then he heard of Dr. Michael Kaplitt.

---

Michael Kaplitt has warded off sleep his entire life. If he could unplug his alarm and set back his OR slot, Kaplitt would crash for 10 or 11 hours a night. But there is too much to do in life. Since he was a kid, he says, he has always chosen to do things that are diametrically opposed to a good night's sleep. At 10 years old, he decided to become a competitive swimmer. While his friends were still tucked in bed, Kaplitt was up at 5 a.m., doing laps before school.

He swam the breaststroke, and after several years of swimming for the YMCA, Kaplitt decided to get serious. Forty miles from his home in Great Neck, Long Island, was the Westchester AAU club. To be a real swimmer, he knew, he had to be on their team. When he was too young to drive, his dad ferried him back and forth in darkness, with Kaplitt often doing his homework by flashlight in the back seat. In high school, he extended the finish line even further. He became captain of the school's swim team, editor of the student magazine, and president of the school. "If you looked at his application to college, you'd say that's what you want your kid to do," says his dad, Dr. Martin Kaplitt.[2]

The muscular six-foot two-inch, 180-pound surgeon appears to be an overachiever—to everyone, that is, but himself. It is as if he sets his sights on something—be it tortuous math problems or the entire Russian language—and simply does it. There is no pomp and circumstance. Life is just a series of tasks to determine and accomplish. It is hard to imagine Kaplitt musing for hours in self-reflection. He is too busy to meditate over his own significance. He does, however, possess a self-awareness of what he is capable of, and what, then, should be expected. There is a sense of duty surrounding him. Kaplitt recognizes his good fortune, and will take every advantage from his place in life.

Kaplitt grew up amidst the estates of Great Neck, a wealthy suburb of New York. It is an exclusive world of luxury cars, stables, and pool cabanas that found fame through F. Scott Fitzgerald's *The Great Gatsby*. His dad, however, was not born with money. Martin Kaplitt spent his childhood living in an apartment in Brooklyn, persevering to one day become a renowned cardiac surgeon. He performed New York State's first operation to clean out a blocked coronary artery, and founded the open heart program at North Shore University Hospital. Michael Kaplitt's grandfather, who became a lawyer, started with nothing. "I always felt you had to do something with your life. You also feel that you're fortunate," says Kaplitt. "There was always a sense that trying to do something with your education made a difference."[3] So he set his sights on an undergraduate biology degree at Princeton University, learning about the science of life.

---

A lot goes on inside the body's 100 trillion cells. There are thought to be approximately 35,000 genes, although cells do not express every gene hiding inside them. With the flip of a biological switch, some genes are turned on while others remain off. While it was first believed that genes themselves control biological function, their power actually lies in what they produce: proteins. Proteins do everything from processing food to combating disease. Every cell has between 10,000 and 20,000 activated genes that churn out proteins.

Genetic diseases can occur when a cell's DNA sequence becomes scrambled, causing faulty genes that produce defective proteins. Because DNA controls all our functioning, the possible effects are almost limitless. Since human life first evolved, people have been at the mercy of their genes. Genetics has been a game of Russian roulette, with our biological destiny left to chance. If gene therapy is as revolutionary as Kaplitt believes, our luck is about to change.

Gene therapy corrects genetic mistakes by inserting new genes into faulty cells. These genes produce healthy proteins that convince cells to return to their normal function. In theory, this is an ideal remedy. If genes can be repaired to churn out the proteins responsible for proper cell processing, the glitch is fixed. But the body is a complex system set up to defend and destroy foreign invaders. How, then, do the new genes reach their target unharmed? By catching a ride on a vehicle called a vector.

As Sheryl Gay Stolberg suggests, vectors are like taxicabs.[4] Viruses are used as vectors to taxi genes to a specific site. The viruses follow the highways and byways of the body, homing in on certain cells while leaving others alone. These spider-like predators latch onto their target cells, sucking out their DNA and replacing it with viral DNA that then replicates and fills the cells with lots of new, strong viruses. This army of predators breaks free from its first victim, killing the host cell and crawling to other surrounding cells to create a tangled, messy web of infection. Some viruses—like the adenovirus, or common cold, which Kaplitt initially studied—affect a person's immune system. In an attempt to defend itself and rid itself of the virus, the body generates an immune reaction. During the internal battle, tissue swells and becomes inflamed. Then there is the classic sore throat and stuffy nose, and typically, in a few days, the virus is sent packing.

Another type of virus is herpes simplex, which attacks neuron (nerve) cells. It is a complex infection that never truly leaves once it has bullied its way through the body's defences. A herpes virus can remain inactive (latent) for months at a time, until the body is under stress and the virus reactivates. The usual result is a cold sore on the lip or under the nose. Once people gain back their strength, the virus recedes, lying in wait for the next opportunity to strike. The class of viruses known as retroviruses are also potential vectors, with the human immunodeficiency virus (HIV) being an example. Then there is the adeno-associated virus (AAV), a single-stranded DNA bundle, and the only virus Kaplitt dares to rely on.

While Kaplitt was an undergrad at Princeton, post-doctoral student Jude Samulski was working down the hall in another lab, purifying and packaging AAV to be used as a gene therapy agent. In 1989, Kaplitt began his Ph.D. in molecular neurobiology at Rockefeller University. His advisor posed a question for him to investigate: How could you insert a gene into the neurons of the brain of an animal to affect its behaviour? Thinking back to his Princeton days, Kaplitt said that he would use a virus. He began researching different viruses, and decided upon herpes as his vector because it naturally infects neurons. Although Kaplitt showed that he could make a virus safe enough to be used in animals, herpes proved to be excruciatingly complex, harbouring a variety of genes that would be difficult to override.

Kaplitt needed to find another vector. He thought of Samulski, and suddenly his "taxicab" was waiting by the curb. Kaplitt, Samulski (then an associate professor at the University of North Carolina, Chapel Hill), and Dr. Matthew During, renowned leader of Yale University's first gene therapy protocol, demonstrated how AAV could be used in the brain. They tested the virus in animals, and it worked. Then came the big test: a Parkinson's animal model. Sure enough, eureka! "Whenever you find something that no one else in the world knows, for that moment, until you report it, it's really quite exhilarating," says Kaplitt. "Even if it's tiny stuff. You feel like you know a little something that no one else knows, except maybe God, euphemistically speaking."

AAV proved to be an ideal vector for gene therapy, in that it had never been associated with human disease. Viruses are born to reproduce and spread. But in order for AAV to complete this life cycle, it relies on another virus to work in tandem. Ideal, if you're a scientist looking to stay on course and not veer into oncoming traffic. But even if a virus is taught to stay on track, it must swerve from many other roadblocks. Say the virus is able to make it to a target cell and crash through its outer membrane "like the wall around a fortress," says Kaplitt. "Once you get in there, you still have to figure out what room the king is in." And depending on the virus, it could come face-to-face with another virus hidden behind a closed door, plotting its own escape.

Take herpes, for example, a virus that remains dormant in the body. Suppose a gene able to kill tumour cells was inserted into a vector designed to attack only these specific cells. Now imagine that, tucked away somewhere in the fortress, there was a dormant, latent virus waiting to be released. If both viruses recombined, the dormant virus could spread with the new gene in it, carrying the gene to other cells and causing a dangerous reaction. While Kaplitt says the chance of this is remote, it is theoretically possible. Except when it comes to AAV. Unlike the herpes vector, which is stripped of only 30 percent or 40 percent of its genes—leaving many genes behind to possibly recombine with latent viruses—AAV is so small that the engineered genes fill up most of its body. And AAV can't create copies of itself, making it scientifically impossible for it to trigger a virus in hiding.[5]

After finishing his Ph.D. in 1993, Kaplitt was ready to move from the laboratory into the clinic, where he could witness the impact of his research. He was having successes in the lab, but none of this mattered if it didn't touch the lives of patients. He enrolled at Cornell University School of Medicine, and today is director of the Center for Stereotactic and Molecular Neurosurgery, and assistant professor of Neurosurgery at Weill Medical College of Cornell University.

Kaplitt remained fascinated by viruses and how they could be used to repair damaged brains. He was drawn to Parkinson's because the disorder threatened life's most vital organ. "The whole body is designed to defend the brain at all costs," he says. "When I talk to my dad and his friends who are heart surgeons, they make a big deal about the heart, and I remind them that the whole purpose of the heart is to pump blood and nutrients to the brain, and that the whole body will shut down in defence of the brain."

And when it came down to testing such an experimental surgery, Kaplitt says, "You're not going to do something cutting edge and science-fictiony for something where there really is no great need. You're not going to do gene therapy for a strep infection when you can treat it with penicillin." In Canada, 100,000 people suffer from Parkinson's disease. This costs the country $558.1 million a year, which includes $39.7 million spent on hospital care, $24.1 million on medication, $23 million on physician care, and $1 million on research.[6]

Parkinson's is the progressive loss of cells in an area of the brain called the substantia nigra. These cells churn out dopamine, a chemical that gives orders to other cells by firing signals throughout the brain. When these neural cells die, dopamine production diminishes, which affects not only the substantia nigra but also other connecting regions. The subthalamic nucleus and the globus pallidus interna become hyperactive, resulting in the typical symptoms of Parkinson's: tremors shooting up an arm or leg, rigidity of the body, and sudden spasms or stutters in movement.

The body goes on autopilot, and its driver becomes a dependent passenger. Once these neural cells die, we know what to expect. The body stiffens, making languid movements impossible. Tremors shoot down arms and legs, rattling limbs uncontrollably. But why this happens remains a mystery. There is a genetic component to the disease, making those with a specific genetic mutation more susceptible, but this is only one factor. It has been suggested that environment also plays a role. Yet, for the majority of cases, physicians still don't know the cause. Most patients try to manage their disease with medication, the most common being L-dopa, which is converted to active dopamine in the brain. But many patients, like Nathan Klein, suffer side effects that are worse than the disease itself. L-dopa can cause dyskinesias, uncontrollable jerking or writhing that contorts the body. When medication fails, surgery is often considered.

Lesioning is a surgical procedure that has been practiced on Parkinson's patients since the 1950s. During lesioning, areas of the brain are heated with an electrode to destroy hyperactive cells. There is also deep brain stimulation, which has, over the past five years, moved from experimental to more common treatment of the disease. In this procedure, an electrode is set within the hyperactive region and permanently implanted. Under the skin, the electrode is connected to a wire that runs behind the head and down the neck to a pulse generator in the chest that looks like a heart pacemaker. Signals are sent by the generator, causing electrical stimulation in the brain that quiets the hyperactive region.[7]

Then there is gene therapy, an experimental and controversial procedure that Kaplitt is audacious enough to bring to the clinic. Rather than using an electric current to quiet the subthalamic nucleus, Kaplitt implants a gene to do the job. In October 2002, after extensive preclinical work in animals, the Food and Drug Administration (FDA) approved the first gene therapy clinical trial for Parkinson's. Kaplitt began a Phase I study at Weill Cornell Medical Center in August 2003 to ensure the safety of the procedure in humans. His first patient: Nathan Klein.

---

Klein proved to be the ideal guinea pig. He met all of the trial's participant requirements. He was under 65 years of age; his disease was no longer manageable by drugs, making him a good candidate for brain surgery in general; and, most crucial in Kaplitt's mind, Klein was psychologically prepared. Kaplitt likes to spend hours with all his patients, trying to convince them out of gene therapy. Instead, he offers to perform deep brain stimulation or lesioning—because if he can place doubt in their minds, Kaplitt says, they are not ready for the trial. The last thing he wants is for patients to have inappropriate expectations that the surgery will change their lives. The trial is designed to test safety, not efficacy.

But Klein felt it was something he had to do. "It's not a matter of bravery. I felt that I had very little choice in the matter. I felt that this was a direction that I was destined to take," he says. "Whether my condition is 20, 30, 50 percent better—or none at all—if this helps figure out how to get rid of Parkinson's, then I've done my job."

"Like some bizarre act of God," Kaplitt says, he found his first patient, who turned out to be a documentary filmmaker. Not only would he be comfortable talking to the press; Klein wanted to shoot footage of his own operation for a future film. Klein's company, TKO Productions, is at work on the project.

Before the cameras started rolling, Kaplitt and his co-principal investigator Dr. Matthew During needed to manufacture the virus. This six-week process was done at During's lab at the University of Auckland, in New Zealand, and at facilities in the United States. The first step is to grow a large amount of human cells that are prone to viral infections in about a hundred plates. Several litres of bacteria are also grown, and harvested for their DNA, which is then inserted into the human cells. The dishes are left alone for a few days to allow the new DNA to naturally begin churning out the virus. The cells are pricked open, and the virus is released. Then comes the purification process. In order to be used in humans, the virus must be extremely pure. All the cellular proteins and debris are sifted out of the future vector. The virus must then undergo several months of sterility testing. Other viruses could have latched on for the ride, bacteria could still be present, or there might be toxicity in the virus.[8]

In order for viruses to be used clinically in the future, Kaplitt says, safety is only one prerequisite. Like any consumer product, the virus must be made consistently, efficiently, and cost-effectively. If the procedure is too expensive, insurers will refuse to cover the cost. If the virus is difficult to handle, surgeons will be more inclined to choose another, simpler procedure. Neurologix—his dad's company, which is funding the Cornell trial—could go out of business, and Kaplitt would have to send his patients home.

Tired of meagre government funding inhibiting research, Kaplitt's dad, Martin, launched Neurologix to market the emerging gene therapy technology. A private equity group structured the company and covered the start-up costs of US$2.5 million. Martin felt the trial was too important to be left to the mercy of the National Institutes of Health (NIH). And he believed their intellectual property held tremendous value. It wasn't the classic case of cashing in on scientific hype. There was an actual product to sell. "If you have something that's important for people, it's going to be valuable financially," says Martin, the president and director of Neurologix. (To avoid conflict of interest, Kaplitt remains off his dad's payroll, working as an unpaid consultant.) But science is infamous for its slow return on investment, and can cost a fortune in the interim. And it could all be for nothing, if the first surgery is a failure.

Nathan Klein was called in a few days before surgery for a PET scan. (PET—positron emission tomography—uses low-energy radiation to create images of brain activity.) Anytime a cell becomes active, it burns glucose (sugar). A radioactive tracer monitoring Klein's brain activity was attached to an intravenous line of glucose, and using the PET scan, Kaplitt zeroed in on active cells in Klein's brain as they ingested large amounts of the radioactive tracer along with the glucose. Klein then went through a battery of neuropsychological tests to gauge his memory, cognitive deficit, and ability to concentrate. An MRI captured images of Klein's brain that Kaplitt used to create a map to lead him to his target. (MRI—magnetic resonance imaging—uses a powerful magnet to scan the brain's structure.) It is a three-dimensional maze of large and small blood vessels and pockets of spinal fluid that, if punctured, could cause the brain to shift.

A few hours before surgery, posts were screwed into Klein's head at four points, two in the front and two at the back of his skull. After the screws had pierced his skin, Klein says, they were tightened so much he thought his skull would crack. A titanium frame was placed over the four posts attached to his head. Then a box was placed over this frame, with markings known as fiducials, which provide a mirror image of what is found inside Klein's head. Using these fiducials, Kaplitt could find any point in the brain relative to its distance from the frame. Then came a CAT scan—computerized axial tomography, which uses X-rays to show a cross-section of the brain—which was fused with the previous MRI scan to create the most accurate brain map possible.

When it was time to begin the surgery, Kaplitt prepared himself. Maybe he took a few deep breaths, or shook the tension from his hands, or maybe he planted his feet solidly onto the OR floor, as if he were back in the starting blocks. Before surgery, he says, "I feel very much the same way that I always felt before a swimming race, which is a little bit nervous, but a little bit excited. But then once you get going, you don't really think about it. You're just in it, and that's it." It was as if the starter fired his gun, and Kaplitt was off.

The surgeon determined his point of entry, on the left side of Klein's skull, where his hairline begins to recede. The area was numbed with local anesthetic only. Using an air-powered drill, Kaplitt made a hole the size of a quarter. Klein says it felt like someone had taken a jackhammer to his head. The frame became a scaffold to guide tubes down the hole into Klein's brain.

With a slow, steady hand, Kaplitt probed neurons in Klein's head with an electrode as fine as a hair. This electrode was wired to a computer, so that the brain's electrical signals could lead Kaplitt to Klein's overactive cells. To keep his cells firing, Klein had to remain awake. General anesthesia would have silenced the target cells and made the trail go cold. There was the soft *rat-tat-tat* from one group of cells, while another set fired off rounds of *click-click-click*. There were spontaneous bursts of noise followed by long pauses.

The electrode snaked its way through the maze of blood vessels, towards the noisiest area of the brain. These were the cells Kaplitt had come for. But the challenge was not yet over. The electrode had to be placed within a millimetre of its mark. If Klein jerked out of position, or if Kaplitt's hand strayed even a hair off target, the outcome could prove deadly. Before the engineered genes could be funnelled into Klein's brain, everything must be aligned perfectly. In some cases, neurosurgeons have to retreat from the patient's skull, reposition, and begin again. This doesn't hurt the brain, but it is extremely time-consuming, which leaves a greater chance for human error.

A nurse wet Klein's lips so that he wouldn't cough. Cautiously, Kaplitt threaded a catheter through the electrode. A syringe was then attached to the catheter. Everything was in place, and the infusion began. Over three billion copies of a gene were taxied through a highway of tubes into Klein's brain to reset the pocket of overactive cells. For an hour and a half,

Klein lay there in one position, his body tender from the cold, steel table, trying to turn himself off, while Kaplitt shook the tension from his aching hands and began to breathe again. Finally it was over. The infusion complete, Klein was released from the tubes and the heavy grip of the titanium frame, and the sterile operating room. He was wheeled into a hospital room, where he stayed for observation for two nights.

While Klein had stopped filming, others began yelling to roll the tape. Klein says he did 25 morning shows, including *Good Morning America* and the *Today* show. The Cornell press conference made Kaplitt and Klein front-page news. But Kaplitt seemed untouched by the hype. "He doesn't care about any of this," says his dad. "He wants to cure disease. Then he'll come back and accept any accolades. Until then, he doesn't want to talk about it."

It is as if Kaplitt has tuned out the stadium of reporters, physicians, and scientists, his eyes fixed on the finish line. He is in a race against himself, testing his own stamina, seeing how long he can go before coming up for air. With a fixed gaze ahead Kaplitt seems to wonder, even before he crosses the finish line, what he will do next. When he pulls himself out of bed every day, Kaplitt searches for the next big thing ". . . to make my life worth something. To some degree it's pathological," he admits. "I looked around after we started this trial, and I thought to myself, What am I going to do after this? Because it's not easy to just follow this. And yet I'm not ready to retire, so it does give you a slight sense of anxiety, and that's probably not entirely healthy, but what can you do?"

After the Phase I study, Neurologix was broke. It cost US$1 million to treat 12 patients, not to mention the running expenses of the laboratory, renting facilities at Cornell, patents, legal fees, and scientists' salaries. Neurologix must drum up a lot more money to fund the Phase II trial, which will cost approximately US$80,000 per patient.[9]

But not everyone is cheering in the stands. Michael Kaplitt's critics feel that he's rash and hasty, that it is too soon to be experimenting with gene therapy, and that Kaplitt is setting himself, and everyone else, up for a fall. Using a virus as a vector could be risky, possibly fatal—for patients and the field of medicine. If a patient dies, along with the patient, so may gene therapy.

---

The mere mention of Jesse Gelsinger can rattle the steady hand of most surgeons. In 1999, the rebellious Arizona teenager found meaning in life. He would volunteer for a gene transfer trial and help save a group of babies who were dying just as they were greeting the world. While Jesse struggled with the usual mood swings and growing pains of adolescence, he also fought a far more personal battle with ornithine transcarbamylase deficiency syndrome (OTC), a rare metabolic disorder that prevents the body from ridding itself of ammonia. The metabolic condition affects one in 40,000 newborns, most of whom are comatose within 72 hours of birth and half of whom die within a month.[10] The teen decided to place himself under a microscope to help these dying babies. Days after his infusion, the 18-year-old became famous through death. Years later, the field of gene therapy is still struggling to release itself from Jesse's memory.

In most ways, Jesse Gelsinger was a normal teenage boy. He liked wrestling and motorcycles, and worked at driving his dad crazy. Paul Gelsinger remembers picking up his son from his after-school job as a supermarket clerk in Tucson one day. On the drive home, they discussed the teen's critical need for medical insurance, and how it might run out when he graduated from high school in a few months. Would the supermarket cover his medication, which cost over US$3,000 a month? Jesse had "forgotten" to look into the matter. He had been sick all his life, taking nearly 50 pills a day, and did everything he could to rebel against his disease.

The frustrated teen threw a bottle of pop against his dad's windshield, cursing his lot in life. Then, a few blocks from home, Jesse threw open the passenger door and jumped out of the van. Paul slammed on the brakes, pinning his son under the right rear tire. Panic took hold, but Paul told himself to remain calm and think rationally. Gently, he rolled the three-tonne van off the teen's arm. Jesse wailed. The pain was torturous, as was Paul's anxiety over his son. Waiting for an ambulance, he cradled Jesse in his arms. "You idiot. What were you thinking?" he yelled, before he broke down and cried, "Jesse, I'm sorry." While Paul says he was an emotional wreck for a week, his son managed to walk away with some road rash and a month of physiotherapy.[11]

In the liver, there are five enzymes that help to expel ammonia from the body. If these enzymes aren't working at full capacity, ammonia—the same substance used to scrub floors—collects in the blood and finds its way to the brain, inducing a coma and eventually death.[12] But Jesse was an anomaly. OTC hadn't commandeered his entire liver, and some cells continued to function normally. With only partial OTC deficiency, he was known as a mosaic, and his disease could be managed with the right combination of medication and diet. When his ammonia did flare up, usually because he had eaten too much protein, Jesse would become bug-eyed and distracted and vomit uncontrollably. By giving him sugary fluids and putting him to bed to sleep it off, his dad said, they could manage the disease.

Still, when Jesse was told of a Phase I gene therapy trial at the University of Pennsylvania (Penn), he was immediately intrigued. The trial was aimed at babies with fatal OTC. At first, Penn intended to study these sick newborns. But Arthur Caplan, director of Penn's Center for Bioethics, convinced the university to monitor adult OTC patients who had their disease under control.[13] Caplan argued that parents of desperately ill children were unable to give proper informed consent for trial participation: "They are coerced by the disease of their child."[14] In other words, they were blinded by the hope of saving their babies, which was not the intention of the trial. Phase I trials are designed to test safety, not efficacy.

From the beginning, this decision proved controversial. Experimental therapies are typically reserved for the gravely ill, because they have little to lose if the treatment is unsuccessful. And in this particular study, the virus intended to taxi the genes through the body was known to cause liver damage, and some scientists worried it would upset ammonia levels in stable adult patients.[15]

Still, Penn set its sights on adults with OTC. A five-hour ammonia test determined that Jesse was a candidate. His liver had 6 percent efficiency in naturally expelling ammonia from the body, and once he turned 18 he was within the age requirement. Even though Jesse knew the study wouldn't help him personally, he was adamant about participating. This made his dad very proud. "This kid has always been a punk. He was a pistol. He could be very selfish, and cruel to his sister. A very normal kid," he says. "And here he was doing this beautiful, perfect thing."[16]

Paul was informed of another OTC patient who had a 50 percent increase in her ability to excrete ammonia after the gene therapy. And he was drawn in by the possibility that one day these sick babies would be saved. "I totally dropped my guard at that point. I didn't question anything any more about it," he says. "We were excited about efficacy. Jesse was going to help them with a breakthrough here. And they were going to save kids. The way that was described to us, there was no way to say no and like yourself."

In the summer of 1999, Jesse was enjoying his young life. He had finished high school on time, despite his health problems. For graduation, Paul bought him a used motorcycle, and he passed his driver's test. He was really living, says his dad. Then, on September 9, Jesse travelled from his hometown of Tucson to Pennsylvania, his first trip by himself. Both excited and nervous, he said goodbye to his dad at the airport.

"Words cannot express how proud I was of this kid. Just eighteen, he was going off to help the world," says Paul. "As I walked him to his gate I gave him a big hug, and as I looked him in the eye, I told him he was my hero." Paul spoke to Jesse that night, and again on September 12, the night before his infusion was scheduled. Jesse was nervous. His ammonia was elevated, and he had been put on intravenous. "I reasoned with him that these guys knew what they were doing," says his dad, "that they knew more about OTC than anybody on the planet."[17]

The next day, Paul was informed that the infusion had gone well. While he was sedated, two catheters had been inserted so that Dr. Steven Raper—Jesse's surgeon and a co-investigator of the clinical trial—could inject 30 millilitres of the adenovirus vector carrying the new genes. A few hours later, Jesse was back in his room. That night Paul talked to

Jesse, who was fighting a fever and the exhausting effects of the virus. They had been prepared for flu-like symptoms and weren't worried. Jesse echoed his dad's words, "I love you too"—the last time Paul would hear these feelings answered.

On September 14, Jesse's ammonia levels started creeping up. Paul had planned on arriving at the hospital for the liver biopsy, which he felt posed the most risk. But with Jesse suddenly in trouble, he came sooner. He arrived on September 15 to find that his son had slipped into a coma. Jesse's ammonia levels were ten times the level of those of a healthy individual. The doctors induced a deeper coma to trigger a ventilator to breathe for Jesse, which seemed to work.

Hurricane Floyd was headed for Philadelphia at the same time that Mickie, Paul's wife, was to land. (Pattie Gelsinger, Jesse's birth mother, had long battled schizophrenia and could not travel to see her son.) Mickie's plane touched down just before the airport was closed. She and Paul went to see Jesse, and could barely recognize him. His young body had become disturbingly bloated; his eyes and ears had swelled shut. Apart from his tattoo and the scar Jesse had from jumping out of the van, Paul felt he was looking at a stranger. Exhausted, Paul and Mickie went to their hotel. For an hour, Paul tossed and turned. The hurricane had retreated to sea and, unable to calm down, Paul walked through the rain to the hospital. He saw blood in Jesse's urine, a sign that his kidneys were failing. "How can anybody survive this?" Paul thought, as he made his way back to the hotel to try to sleep.

On September 17, Paul and Mickie were told that Jesse's organs were shutting down, and that he had suffered irreversible brain damage. Knowing it was all over, Paul gave the order to turn off Jesse's ventilator. A chaplain was called in. Suddenly, it was as if the air had been sucked from the room. Dr. Raper lifted Paul's hand from his son's chest, replacing it with a stethoscope. The physician noted the time of death and said, "Goodbye Jesse. We'll figure this out."[18]

"Within four days of treatment, he's dead. And I had to shut off life-support," says Paul. "At moments I had real anger towards [the Penn medical team], but what Jesse had demonstrated for me was overwhelming, and that's the perspective I was taking."[19] Six weeks later, family and friends carried Jesse's medicine bottles filled with his ashes to the top of Mount Wrightson, south of Tucson, to say their final goodbyes. It had been one of Jesse's favourite places to be.

The next few weeks were very emotional and confusing for Paul Gelsinger. He met with Dr. Robert Erickson, a University of Arizona medical geneticist who had reviewed the OTC trial in 1995. With Raper in the room, Erickson explained his apprehensions about the study, since several monkeys had died in preclinical work. Later, when Raper was alone with Paul, he admitted that, yes, the monkeys had died, but the virus had been adjusted to reduce the toxicity and ensure its safety. This seemed to make sense, and Paul continued to stand behind Penn.[20]

Then, in December, grief was again taken over by anger. Paul attended an NIH meeting to discuss Jesse's death, the first fatality directly linked to gene therapy. Also in attendance were principal medical investigators Dr. Raper and Dr. Mark Batshaw, the pediatrician who had conceived of the clinical trial, as well as Dr. James Wilson, co-principal investigator, who was also director of Penn's Institute for Human Gene Therapy. The physicians were asked to describe the trial to the NIH's Recombinant DNA Advisory Committee (RAC), which monitors gene therapy studies. Each of the 18 adult participants had received an infusion of the OTC gene into the liver using an adenovirus vector. Jesse had been given 38 trillion virus particles, the trial's largest dose. Only 1 percent of the genes had reached the liver cells. None of the patients had illustrated significant gene expression[21]—that is, none had shown significant improvement in handling ammonia.

Utter confusion washed over Paul. This made no sense. What about the patient he had been told about, who could secrete ammonia 50 percent more efficiently than before she had the gene transfer? "I'm sitting there thinking, what do you mean, no efficacy?" Paul asked himself. "You guys. . . . I was told this thing was working."

Paul was blindsided by what he learned. The FDA, the arbiter of all U.S. clinical trials, had not been informed of a change in the wording of the participant consent form, which left patients unaware of the death of several monkeys given a similar dose.[22] The FDA had not been notified that participants had suffered side effects so severe that the trial should have been halted before Jesse was treated. Paul also discovered that, according to the FDA, Jesse should never have been allowed to participate when he had such high ammonia levels.[23] After the NIH meeting, Paul says, "I left there and I cried for an hour. And I couldn't believe it. I said, Oh my God, I've been used. They were using my support for them to cover their tracks."

During the months after the clinical trial, the medical investigators denied responsibility. Shortly after the trial, in an interview with *The New York Times Magazine,* Dr. Wilson admitted his fear was that "I'm going to get timid, that I'll get risk averse." Dr. Raper told the *Times* that he would live up to his promise to "figure this out." Dr. Batshaw seemed to be taking Jesse's death the hardest. He struggled with his role in the teen's death, and asked, "What is the Hippocratic oath? . . . I did harm."[24]

So thought Alan Milstein, the lawyer whom Paul hired after attending the RAC meeting. Milstein, who has since made his name as a crusader against the abuse of humans in clinical experiments, says, "The history of medicine is replete with instances of progress at the price of the dignity of human subjects. Greater good versus the compromise of the individual."[25] Milstein points to the Nuremberg trials in which Nazi doctors were convicted of abuse during the Second World War, and the human radiation experiments carried out by the U.S. government during the Cold War.

---

In 1996, the International Conference on Harmonisation (ICH) created the "Guideline for Good Clinical Practice" for clinical trials performed in the European Union (EU), Japan, the United States, Australia, Canada, and the non-EU Nordic countries. The guidelines state, "Before a trial is initiated, foreseeable risks and inconveniences should be weighed against the anticipated benefit for the individual trial subject and society. A trial should be initiated and continued only if the anticipated benefits justify the risks. The rights, safety, and well-being of the trial subjects are the most important considerations and should prevail over interests of science and society."[26] If a gravely ill patient who has exhausted other possible treatment and therapy participates in a clinical trial, the benefits for society seem to outweigh the risks to the individual. If the person is unlikely to live, how can he or she be taken advantage of?

As for the OTC trial, Milstein says, "The design of this experiment was such that it never should have started. The risks outweighed the benefits." The chosen participants were not gravely ill. In fact, it was a distinct possibility that their condition could worsen because of the experiment. According to the University of Vermont's Community Genetics and Ethics Project, "in this trial the healthiest population was enrolled as volunteers, those who were able to manage their disease through conventional treatment. Because these patients' livers are already stressed due to their disease, it was very risky to treat them with a therapy proven to cause liver damage. Since the volunteers stood no chance of any lasting benefit from the therapy, the risks far outweighed the benefits in this trial."[27]

In general, says Milstein, there is an ethical dilemma with Phase I trials, in which researchers are testing the safety rather than the efficacy of a therapy. Toxicity tests are not expected to offer therapeutic benefit to patients, even though participants and their families hope for just that. "There's no benefit to the subjects. And since the purpose of the study is to find the point of harm, you're talking about all risk and no benefit," says Milstein. "If you're controlling it to the point where the risks are relatively minimal, then maybe you can ethically do the study. But if you can't, then you've got a real ethical problem on your hands."

Then there is the moral dilemma of informed consent. "We know that informed consent doesn't work," says Arthur Caplan. The privatization of science and conflicts of interest, along with the fact that patients are often blinded by the hope of a cure, he says, prevent true informed consent.[28] Yet written consent from all participants is a trial requirement. Patients must understand the benefits and risks of the trial. And according to the ICH's guidelines, the patient consent form must be revised whenever important new information becomes available that could affect a person's consent.[29]

If adverse reactions do occur in U.S. trials, they must be reported to the FDA and the NIH. But the rules differ among these organizations. The NIH's RAC concerns itself only with gene therapy trials that the NIH is helping to fund. While adverse events must be reported, some confusion arose after the RAC was reorganized in 1997, at which time its authority to approve gene therapy trials was handed over to the FDA.[30] After Jesse's death, the NIH sent out a reminder that all adverse events must be reported. To its absolute shock, and to the dismay of the entire medical community, officials were hit with 652 new reports of adverse events from about 80 institutions. While Jesse was the first to die from gene therapy, it was revealed that at least seven earlier deaths had gone unreported.[31] Because of the 1997 changes within the RAC, researchers may have felt their duty was to report adverse reactions to the FDA, and not to the NIH.

All U.S. trials, regardless of funding, must follow FDA guidelines, which are similar to those of other countries. Health Canada requires that adverse events be reported within 15 days, if they are directly related to the therapy or if they are serious and unexpected.[32] Yet, unlike RAC reports, these events are never made public. If the FDA or Health Canada decides that a trial is unsafe, it has the power to shut it down. But no one outside these government organizations is privy to such sensitive information. Not researchers in the field, who could alter their own trials accordingly. Not potential future participants, who might opt out if they knew the risks in similar studies. Such health information, by law, is confidential, says Dr. Anthony Ridgway of Health Canada, who wrote "Regulation of Gene Therapy: The Canadian Approach." He says this is intended to protect proprietary information that is often related to the manufacturer or to special aspects of the product itself. "It's outrageous," says Milstein, "that adverse effects are proprietary to the company that's performing the research."

In 1995, the FDA and NIH were in the process of implementing the Gene Therapy Information Network (GTIN), a publicly accessible electronic database of results of gene therapy trials, which would include adverse reactions. But the project was dropped after a heated RAC meeting involving RAC committee members and FDA representatives. Paul Gelsinger was tipped off about the meeting. He found the minutes on the RAC website, and was appalled by what he discovered. The FDA wanted the database to go away. RAC representatives demanded an explanation for cancelling the database. Dr. Philip Noguchi, director of Cellular and Gene Therapies for the FDA, was quoted as saying that his superiors answered to industry. Alexander Capron, a RAC committee member, said he thought the FDA answered to the American people. [33]

In 2002, American and Canadian industry spent US$2.9 billion on academic research, compared to the US$23.1 billion spent by federal government sources, according to the Association of University Technology Managers. In 2002, academic research generated 450 new companies, with academic institutions receiving an equity interest in 69.6 percent of their start-up companies.[34]

Since the Bayh-Dole Act[35] was passed in 1980 to bolster American industrial production, by enabling universities to patent federally funded research, the relationship between academia and industry has grown cosier and cosier. Before Bayh-Dole, universities secured around 250 patents annually; in 1998, eight years after the legislation was enacted, there were more than 4,800 patents secured.[36] From the beginning, people had been divided over the act. Proponents of academia said Bayh-Dole would turn ivory-tower ideas into commercially viable products for public use; it would fund the academic mission of universities, enhance regional economic development, and offer more employment opportunities for students. Proponents of industry said corporations would become forerunners of innovation, academic research would help expand companies' intellectual property, and universities could facilitate contracts for the corporate laboratory.[37]

The benefits of such academic-industry collaborations have not gone unnoticed by government. "In a global environment in which prospects for economic growth now depend importantly on a country's capacity to develop and apply new technologies," noted Federal Reserve Chairman Alan Greenspan, "our universities are envied around the world. The payoffs—in terms of the flow of expertise, new products and start-up companies, for example—have been impressive."[38]

Japan decided to cash in on its own industry-academic collaborations. In 1995, the government's Science and Technology Basic Law allocated 17 trillion yen for scientific research at national universities over five years. In 1998, the Technology Licensing Organisation (TLO) Promotion Law was launched to encourage the transfer of technology from academia to industry. Since 2000, the Industrial Technology Reinforcement Law has enabled university faculty to become directors of industry. And in 2002, the Intellectual Property Basic Law was created to help industry, universities, and government in Japan work together.[39]

Yet the fear is that money will dictate science, that a researcher's financial interests will impede the free, unbiased pursuit of knowledge. What if a discovery could make a scientist a millionaire through stock options? What if disclosure of negative results compels a company to pull its funding from a university, slashing research budgets and forcing laboratories to close? The World Medical Association Declaration of Helsinki, "Ethical Principles for Medical Research Involving Humans," was promulgated to provide guidance to physicians around the world. The declaration states, "In any research on human beings, each potential subject must be adequately informed of the aims, methods, sources of funding, any possible conflicts of interest, institutional affiliations of the researcher, the anticipated benefits and potential risks of the study and the discomfort it may entail."[40]

If a researcher earns an annual salary of $10,000, receives $10,000 from equity holdings, or has 5 percent ownership in a company, a 1995 U.S. federal government regulation says this information must be disclosed to universities.[41] According to the American Association of University Professors, ". . . financial ties of researchers or their institutions to industry may exert improper pressure on the design and outcome of research."[42] To address the issue, universities have created their own research ethics boards (REBs) that oversee clinical trials and ensure that there is no conflict of interest. Canada's Medical Research Council, Natural Sciences and Engineering Research Council, and Social Sciences and Humanities Research Council state in the "Tri-Council Policy Statement: Ethical Conduct for Research Involving Humans," "When a significant, real or apparent conflict of interest is brought to its attention, the REB should require the researcher to disclose this conflict to the prospective subjects during the free and informed consent process."

For REBs to determine conflicts of interest properly, the policy statement says, they should be given details on research budgets, commercial interests, and consultative relationships. It goes on to state, "Sometimes, the conflict of interest is so pervasive that it is not enough merely to disclose it to the research subjects, the sponsors of research, institutions, relevant professional bodies or the public at large. In such instances the REB may require that the researcher abandon one of the interests of conflict."[43]

Gene therapy first entered the clinic in 1990. Since then, it has been a game of catch-up for ethics boards, government bodies, and researchers themselves. Over the past 15 years there has been a process of trial and error to determine what regulations are needed for this unique and experimental field of medicine. How, for example, is gene therapy similar to and distinct from pharmaceuticals? Health Canada regulates gene therapy as it would any biological drug, says Ridgway, adding that there are no specific guidelines created for gene therapy. While he hopes Health Canada creates such regulations one day, Ridgway says it is a daunting task. "Each and every vector has its own attendant concerns, potential problems, and potential adverse events," he says. "So it gets fairly complicated to come up with any guidance that is truly comprehensive."

---

Dr. W. French Anderson, the physician who first brought gene therapy to the clinic, was well aware of its complications. In 1990, his concern was that "we might be like the young boy who loves to take things apart. He is bright enough to disassemble a watch, and maybe even bright enough to get it back together again so that it works. But what if he tries to 'improve' it? Maybe put on bigger hands so that time can be read more easily. But if the hands are too heavy for the mechanism, the watch will run slowly, erratically, or not at all."[44]

Anderson, known around the world as the father of gene therapy, could have been that boy he spoke of who pulled things apart to peer inside and figure out what made them tick. He grew up in Tulsa, Oklahoma. From a young age, he was intent on learning how the world worked. At eight years old, he pored over college science books. Even though an incessant stutter made him the target of ridicule, Anderson was unflappable, "because I considered everybody else in the world stupid."[45] He overcame his speech impediment by talking with his cheeks full of pebbles, and later forced himself to join his high school debating team.

Anderson earned an M.D. from Harvard Medical School in 1963, and joined the NIH. He stayed with the institute for 27 years, intent on discovering how to use genes to correct disease. In 1968 he was already predicting that the first attempts to correct genetic defects were only years away.[46] But the visionary doctor would have to wait two decades before the experimental therapy was lifted from the realm of dream to medical reality. In 1990 Anderson got approval for the first gene therapy trial, on four-year-old Ashanti de Silva, who has ADA-deficiency—a rare disorder that doesn't allow her to produce adenosine deaminase, an enzyme integral in maintaining a healthy immune system. Between 1990 and 1992, de Silva had 11 infusions of gene-corrected mature T-cells.[47] (T-cells are a form of white blood cell involved in the immune system.) The trial was declared a success, and gene therapy was an overnight sensation.

If Anderson is the father of gene therapy, Dr. James Wilson could be called the son. An M.D. and Ph.D. from the University of Michigan, Wilson joined the prestigious Whitehead Institute of the Massachusetts Institute of Technology in the 1980s, where he focused on retroviral vectors and gene therapy. In the 1990s, when science and business leaders were proclaiming that gene therapy was the next big thing, they looked to Wilson to lead them into the future. He became director of the University of Pennsylvania Medical Center's renowned Institute for Human Gene

Therapy, and was often profiled by highly respected science journals.

Milstein is the first to say that Wilson was Anderson's heir apparent. "He was by all accounts brilliant, by all accounts dedicated, hard-working," says the Gelsingers' lawyer. "He truly wanted to bring this field into a place where it actually accomplished something as opposed to being theoretical. He wanted to be the next Jonas Salk [inventor of a vaccine for polio]. And he wanted to save the babies of the world, really." Wilson's credentials were never called into question. Rather, it was his motivations.

In the early 1990s, Wilson decided to place his own bet on the emerging field of gene therapy. He took out a second mortgage on his house and launched Genovo Inc. The firm paid his laboratory at the University of Pennsylvania US$2.8 million annually. His university students worked on research funded by Genovo. And the university itself was a shareholder, ready to cash in on royalties. According to *Forbes,* Wilson was so intertwined with industry that the University of Pennsylvania set up a conflict-of-interest committee to oversee the Genovo alliance.[48] The 1999 article states, "His double life draws the envy of some colleagues and the ire of others, who worry about where his loyalties must lie and fret that the pursuit of truth has been corrupted by the pursuit of royalties."[49]

After Jesse Gelsinger died, Wilson's judgement was called into question. What had been his motivation for going ahead with the study despite the fact that several monkeys had died during testing, that participants had suffered adverse reactions, and that Jesse's ammonia levels had become elevated? Had it been the pursuit of stock options rather than the pursuit of knowledge? In 1995, Biogen Inc. had paid Genovo US$37 million to market any future liver-related therapies that arose from Genovo's research.[50]

Paul Gelsinger says he was never told of Wilson's commercial interests in his son's trial. Several weeks after Jesse died, Paul invited Wilson to his house for dinner. The two men spent hours together, hiking around Tucson. Paul was consumed by the FDA representative's comment that his superiors answered to industry. Was the Penn team kowtowing to commerce? "I asked him, What's your financial stake in this?" says Paul. "Oh yeah, I asked him to his face. Sitting on my back porch, he said, I'm an unpaid consultant for the biotech company once a week." After Jesse died and the Penn trial was halted, Genovo was sold. Wilson kept stock options estimated at US$13.5 million, and Penn held onto equity of US$1.4 million.[51]

"Everybody was being influenced by the money. And to me, the ultimate culprit here is the money," says Paul. "It was what made these men lose sight." While Milstein agrees, he believes "it wasn't that the researchers compromised their values for the almighty dollar." For Wilson, there was much more at stake than a fat cheque. Great things were expected of him, as director of Penn's Institute for Human Gene Therapy, and Jesse Gelsinger could make him a medical legend. The fact that Jesse was a mosaic, with some of his liver cells functioning while others didn't, made him "ideal for research," says Milstein. "It would be like being able to experiment on the babies." Not only would Wilson have proven safety, he would also have shown the lab coats and suits of the biotech world that this method actually worked.

It is Milstein's view that the Penn team set their sights on Jesse years earlier: "They wanted Jesse Gelsinger." For Paul Gelsinger it is painfully ironic, looking back now to the place where Jesse was first diagnosed: Children's Hospital, adjacent to Penn's medical facility. Jesse became famous for his rare form of OTC. He was even written up in *The New England Journal of Medicine*.

The case of Jesse Gelsinger never made it to court. On November 3, 2000, the University of Pennsylvania settled with the Gelsinger family. The details remain undisclosed. The fallout: all human trials were suspended at Penn's Institute for Human Gene Therapy. Rather than comply with FDA changes, the university announced that it would no longer conduct human studies there.[52] Penn admitted to the FDA that it should have ended the trial once adverse effects were shown in patients. But it disagreed that Jesse was ineligible for the study, and denied liability for the 18-year-old's death.[53]

In a letter to Wilson dated February 8, 2002, the FDA stated, "The Food and Drug Administration (FDA) has information indicating that you repeatedly and deliberately violated federal regulations in your capacity as investigator in clinical trials with unlicensed biological and investigational new drugs, specifically, an adenoviral vector. These violations provide the basis for the withdrawal of your eligibility as a clinical investigator to receive investigational new drugs."[54] In July of that year, Wilson stepped down as a director at the University of Pennsylvania.[55] (He continues to work in the university's division of medical genetics.) The Institute for Human Gene Therapy was disbanded.

Much has changed because of Jesse. The NIH contacted every gene therapy clinical investigator to ensure that all serious adverse effects are reported, including those for past trials; the RAC/FDA web database finally got up and running; and the FDA and NIH perform regular inspections to check that trial investigators are complying with their protocols.[56]

Gene therapy will forever be associated with Jesse Gelsinger. But maybe one day it will become famous for eradicating genetic diseases like cancer and Parkinson's. If viruses are perfected to taxi genes to sick organs, if those genes are properly programmed to reset cells, then the body may be rewritten, and genetic diseases erased from our lives. Until then—as Dr. Michael Kaplitt quoted Clint Eastwood, in his medical school yearbook—"A man's got to know his limitations."

## Chapter Two

# The Gene Hunters

## On the Trail of Deadly Disease Genes

Nancy Wexler is painfully aware of the limitations of medicine. For more than a decade, she watched helplessly as Huntington's disease held her mother's body and mind hostage. At first, Leonore Wexler lost control of the small things in life. Her feet began to drag; her fingers would twitch without warning. Then it was as if someone had broken into her house and kidnapped her freedom overnight. The most mundane actions became hopeless. Lifting a spoon to her mouth was maddening, as was trying to communicate through slurred, incoherent speech.[1] A biologist, Leonore knew her fate, and made an unsuccessful attempt at ending her life by overdosing on sleeping pills. She spent her last days in a padded hospital bed, after uncontrollable thrashing left her battered and bruised. When she finally died on Mother's Day, 1978, Nancy Wexler says, Leonore "looked like an inmate at Dachau."[2]

It is hard to fathom that such crippling diseases can result from the slightest biological misprint in the body's 100 trillion cells. Each cell contains a nucleus which itself contains two sets of chromosomes, one set descending from the mother and the other from the father. It was the Austrian monk Gregor Mendel who allowed us to blame our genetic makeup on our parents. In 1865, after performing countless breeding experiments with pea plants in his monastery's garden, Mendel established the law of heredity. When he crossed plants bred to grow tall with plants bred to grow short, the next generation of plants sprouted like its tall parent. When Mendel interbred this next generation, both tall and short plants were produced, illustrating that the tall characteristic was dominant and the short characteristic recessive.[3] If one parent has two tall genes (TT) and the other has two short genes (Tt), offspring (tt) will all be tall, because the tall (T) gene is dominant. But if a hybrid tall (Tt) pairs with another hybrid tall (Tt), both short and tall are generated, depending on what combination of genes is passed on. Any offspring receiving a tt combination will be short.

Mendelian inheritance was linked to chromosomes, but at first it was deduced that there were a total of 48 in humans.[4] Since 1956, however, it has been well established that a human cell consists of 46 chromosomes divided into 23 pairs. Every chromosome is composed of deoxyribonucleic acid (DNA).

It was Canadian-born Dr. Oswald Avery who, along with colleagues at the Rockefeller Institute in New York City, concluded that DNA determines a person's biological fate. Using the "transforming principle," introduced by Dr. Fred Griffith, which shows that bacteria can change from one type to another, Avery observed bacteria reproducing. When he destroyed the bacteria's DNA, there was no transformation. In 1944, Avery announced that DNA was in fact the instruction book to life.

DNA's instructions are written out in an alphabet consisting of four letters, A, T, C, G, which stand for the molecular components adenine, thymine, cytosine, and guanine. The order in which letters are written in our personal DNA manual not only determines our appearance, it also influences our health. It sketches our hair and eye colour, and decides which, if any, genetic diseases we are susceptible to. James Watson and Francis Crick, now considered the fathers of DNA research, understood these components. But how, they wondered, did they all fit together to form DNA's intricate structure?

On February 28, 1953, James Watson was in his laboratory at Cambridge University. It was a Saturday, and he was taking advantage of the rare quietness. He had made himself three-dimensional cardboard cut-outs of DNA's bases, and methodically tried to fit together the puzzle of As, Ts, Cs, and Gs. It didn't take long to discover that As seemed to lock with Ts, and Cs with Gs. "It was so simple, so elegant, that it almost had to be right," says Watson.[5] Once Crick verified his findings, they knew they had it: the "double helix," the interlocking structure that was the key to genetics. These two spiral staircases of paired bases, when broken apart, partner with other matching strands to pair bases and make more identical DNA molecules.

While DNA is the manual for storing genetic information, it doesn't itself carry out the instructions. That job is left to the body's proteins. There are structural proteins—like collagen, which help make bone and skin—and functional proteins—like antibodies, which combat infection. To build a protein, a DNA helix is unravelled, leaving two strings of bases. One string acts as a template for ribonucleic acid (RNA), whose job it is to assemble proteins. The RNA floats off, and is translated into amino acids that fit together to make protein.[6]

Genetic diseases occur when genes mutate, or change their DNA sequence, and the faulty genes produce defective proteins.[7] Genetic variations often have nothing to do with a particular gene's normal function; many people live normal lives despite DNA changes. These can be insignificant blips in the person's DNA that he or she will likely never know, or need to know, about.

But there are also genetic mutations that cause disease. There are two types of genetic diseases: genetically complex, and single gene. The first category includes complex diseases like cancer, which result from a combination of genetic susceptibility and environmental factors. While having, say, the BRCA1 gene makes a woman more susceptible to breast cancer, the onset of the disease is also affected by the environment in which she lives. Is she exposed to certain chemicals? How clean is the air she breathes? Does she smoke?

While scientists are certain that both lifestyle and genes influence the onset of complex gene diseases, they still don't know how environmental factors react with certain genes, and how big a part genes play in determining people's susceptibility and resistance to disease. Some people are able to smoke for years and remain unaffected, while others develop lung cancer. There are those who are sensitive to salt and are burdened with high blood pressure, and others who eat as many salty french fries as they please. This can be chalked up, at least in part, to our genes.[8]

There are approximately six thousand single-gene diseases, passed from generation to generation through a genetic mutation. For cystic fibrosis to appear, both parents must pass on the disease-causing mutation. If only one parent has the faulty gene, the child will not have CF but may be a carrier, passing the gene to the next generation. For Huntington's disease, only one parent must pass on the mutation, which is in the gene encoding the huntington protein. This gene is dominant, so it overrides the expression of a normal gene, from the other parent, on the chromosome.

Even before people with the Huntington's defect are born, their fate is sealed. They have been given a death sentence along with the mutation. But their genes dictate how long it will take for the sentence to be carried out. To avoid the disease, there must be no more than 35 repeats of the three DNA bases CAG, with most people having 10 to 15 repeats. "Your destiny, your sanity and your life hang by the thread of this repetition," explains science author Matt Ridley.[9] The number of repetitions also determines the timing of the onset of disease. If there are 39 repetitions, a person will most likely experience initial symptoms at 66 years of age. If there are 42 repetitions, average onset will be at 37. And people with 50 repetitions will typically start to lose their grasp on life at about 27.[10]

Nancy Wexler, who is now the Higgins Professor of Neuropsychology at Columbia University, was in her early 20s when she discovered that her mother carried the genetic mutation for Huntington's. It was 1968, and her father, scientist Milton Wexler, launched the Hereditary Disease Foundation. The goal was to explore Huntington's as well as other debilitating genetic diseases, with the end goal of locating the disease gene.

In 1979, at a NIH genetics workshop, the idea of restriction fragment length polymorphisms (RFLP) markers was proposed. RFLP mapping is, in theory, quite simple. Everyone has unique DNA markers, and following their trail reveals what chromosomes were inherited, and from which parent. It is likely that disease genes are inherited along with these neighbouring markers, whose locations are already known. Using this technique, scientists could then predict many aspects of a life even before a baby has taken his or her first breath.[11]

To find the Huntington's gene, why not look for RFLPs along DNA in families that carry the disease? By doing so, it could be determined whether RFLPs are passed from generation to generation with the gene. This could cinch together the borders around the vast genetic landscape. Imagine, says Wexler, culling through a person's genome, which is long enough to circle the globe. On this scale, the DNA in a chromosome would reach thousands of kilometres; a gene would extend only one-twelfth of a kilometre; and a disease-causing point mutation—an alteration of merely one base pair—could be as brief as one-seventh of a centimetre.[12]

When Wexler began her work with RFLPs, the scientific community balked at this type of genetic mapping. Finding a gene using this method could take upwards of 50 years, critics estimated. Wexler understood their hesitation. "What we were proposing," she says, "was equivalent to looking for a killer somewhere in the United States with a map virtually devoid of landmarks—no states, cities, towns, rivers, or mountains, and certainly no street addresses or zip codes—with absolutely no points of demarcation by which to locate the murderer."[13] To have any hope of finding the Huntington's gene and its markers, Wexler knew, they needed a large population with the disease. Soon after the NIH workshop, Wexler found her family.

---

Pull back the curtain of Venezuela's dense jungle and you will find Lake Maracaibo, one of the world's richest oil reserves. Shifting rainbows of oil glide along the water's surface. Oil towers line the shore, along with piles of forgotten garbage. Among the pipes churning out the country's richest resource sit the isolated villages of San Luis, Barranquitas, and Laguneta. Their people have a secret, and it is killing them. They have the world's highest incidence of Huntington's disease. In Barranquitas, nearly half the 10,000 residents carry the disease-causing gene. Since Venezuelan doctor Americo Negrette first recognized his country's predisposition to the disease in the 1950s, the number of those afflicted has surpassed

16,000, most of whom can trace their inheritance back to the 1800s, to a woman aptly named Maria Concepción.[14]

Inbreeding is normal here, as is having 12 to 14 children, which gives Huntington's a passport to immortality. Villagers live in poverty, fighting the disease in the confines of tin shacks. Most children have never seen the inside of their local school; they must stay home to look after their dying relatives. The men make their living fishing, venturing into the dirty waters at night. When the disease takes hold, they begin falling overboard. Some drink gasoline to end their suffering, or attack each other with knives. Men often abandon their sick wives.[15] For these villagers the question isn't if, but when, the disease will steal their lives.

Americo Negrette introduced Wexler to the villagers of Lake Maracaibo. In 1981, Wexler made the trek from her American home to Venezuela with a team of researchers. To study Huntington's, she needed blood samples of family members with and without symptoms of the disease. Most villagers had never given blood before. "Some men believed that if they gave blood, they wouldn't be able to drink, and they didn't want their drinking interrupted. So we had to resort on occasion to drawing blood from one arm and giving them a beer in the other," she says. "The hardest people to collect from were the old and healthy. They had escaped the illness. They were scared because we wanted their blood too. What did that mean?"[16] But Wexler showed them her scar from a skin biopsy, which gave her credibility. The villagers decided she could be trusted. "I and my mark became something of a passport for our research team and its activities."[17]

Blood samples could be drawn only when someone was leaving Lake Maracaibo and could hand-carry the vials safely back to the United States, to molecular geneticist James Gusella's laboratory at the Massachusetts General Hospital in Boston. Wexler said these Venezuelan days, known as "draw days," were "chaotic days in boiling hot, deafeningly noisy rooms jam-packed with people of all ages, days spent going to sweltering homes where throngs of children would shout out in gleeful horror the number of tubes of blood we were drawing."[18]

In Boston, Gusella sifted through reams of DNA for a genetic marker that would bring him closer to the Huntington's gene. He also used samples from a large family in Iowa. He cut up pieces of DNA using restriction enzymes, making RFLPs—in other words, genetic markers—that he labelled using a radioactive tracer. These markers had many forms, so individuals could be distinguished from one another.

Pieces of DNA were placed on a gel that separates fragments by their size. Then Gusella added the radioactive marker, and when it met its DNA counterpart he had a match. The markers would light up, creating unique bands that could be traced to either symptomatic or asymptomatic family members. If those who were already sick all had the same radioactive bands on their DNA, there was a good chance the Huntington's gene was on the same chromosome.

After testing only three markers, Gusella got lucky. The marker was called G8, and the odds were better than 1,000 to one that it was near the Huntington's gene. Those already fighting the disease had one form of the marker, while those still in good health had another. The discovery was announced to the world in the November 17, 1983 issue of *Nature*. Gusella, Wexler, and their team were in awe, as were their critics. In three years, they had narrowed Huntington's to chromosome four. Everyone agreed that luck was on their side.

Yet for Gusella, success was bittersweet. He was haunted by what the discovery meant for the sick: nothing. He had found a marker predicting an incurable disease, but what did this really mean if it couldn't lead to treatment? "A healthy person, thus, could learn that in 15 years they're going to look like their sick parent and there's no hope. That's a hell of a quandary, let me tell you!" he says. "It put tremendous pressure on all of us to find the gene because if we know what the defect is, we have at least the possibility of developing some kind of rational therapy."[19]

The hunt was on. And Wexler, who is often referred to as the cheerleader for a Huntington's cure, was full of hope. Out of three billion possible base pairs on 23 chromosomes, they were a mere four million base pairs away from their target.[20] But they were also at a crossroads. Like tourists in a foreign city trying to decipher signs in an unfamiliar language, they didn't know which way to turn. They deduced that the gene was most likely on the tip of chromosome four, a mystifying region of about 150,000 base pairs that Wexler calls "the Twilight Zone of genetics."

For years the team attempted to find markers on both sides of the gene, to no avail. Without biological signs as to what the gene was doing, they were unable to track it by working backwards. Then, in 1990, disaster struck. Gillian Bates and her team at London's Imperial Cancer Research Fund were studying the tip of the chromosome to see if the gene was hiding there. It suddenly became apparent that Wexler and her group were in the wrong place. They had turned the wrong way. It was two million nucleotide links in the opposite direction.[21]

There was no choice but to persevere, which they did for many years. Then one day Marcy MacDonald, one of Gusella's senior researchers, found a repetition of the bases cytosine, adenine, and guanine in the targeted area. When she and Gusella compared the number of repeats in healthy people to repeats in those showing signs of the disease, the latter always had more repeats. Could this be what they had been tirelessly pursuing for more than a decade? Did it all come down to a simple repetition, a point mutation hiding in a field of bases? Again and again they tested the repeat, in sick individuals and healthy ones. The healthy people showed between 11 and 34 repeats, while those who were sick showed between 37 and 86.[22]

Another secret was uncovered: the more repeats a person had, the sooner he or she would show signs of the disease. They tested a boy who fell ill at three years old. He had 86 repeats, the most they had seen. Their findings were conclusive, and their quest a triumph. The results were published in the February 26, 1993 issue of *Cell,* with Wexler's name among the collaborators.

When Gusella let Wexler in on the news, she couldn't stop jumping up and down, with tears flowing freely.[23] She found her way back to Lake Maracaibo to tell her second family the joyous news. "I had spent so many years being so curious about what it was, studying all these people whose bodies contained the mystery," she says. "And suddenly it was superimposed on them, almost like a silk screen. It was an image without words, saying, Here's the answer. And here's another question."[24]

The most difficult decision for Wexler was whether to get tested herself. Did she carry the faulty gene? With how many repeats? As with villagers of Lake Maracaibo, a simple blood test would reveal her future. Sometimes, everyday accidents felt like clues to her destiny. Most of us think nothing when something slips from our grasp and falls to the ground, or when we trip while walking down the street. We chalk it up to our being preoccupied or careless. For Wexler, such mishaps are terrifying, because any one could signal the beginning of the end. For over 30 years, she has wondered whether she will be spared. But she has chosen to take her cues from life rather than from a genetic test.

---

Wexler and her sister, Alice, have decided never to have children. Whatever happens to them, they choose not to pass the faulty gene on to another generation. But some people feel the decision should not be theirs to make. The thought of that choice being taken from them is a frightening one.

It was Charles Darwin's cousin Sir Francis Galton who proposed the idea of eugenics—of sterilizing those deemed genetically impaired. Sir Francis chose the word from a Greek root meaning "good in birth" or "noble in heredity." The goal: to rid the world of the unwanted, the embarrassments of society. It was the "feeble-minded" that the eugenicists set their sights on. They were seen as carriers of crime, poverty, and disease. Strip them of their ability to pass along genetic misfortunes, and these problems would diminish.

A report by Richard Dugdale, a New York prison inspector, proposed eugenics to the public. Dugdale observed that a large number of inmates at an upstate prison were relatives. He compiled a family tree of 709 people from five generations who could be traced back to a Dutch settler. These individuals had "a propensity for almshouses, taverns, and brothels."[25] This new type of sociological study was the first of many. During the mass immigration of the 1900s, many Americans began to consider the possibility of eugenics.

Immigration spiked from 225,000 in 1898 to 1,300,000 in 1907.[26]

Buildings erected across the United States to house the overflowing population quickly became slums. Competition for work was fierce, and many Americans blamed immigrants for this new reality. Their American dream was being infringed upon not by other true Americans, but by an influx of Dutch, Italians, Jews, and other foreigners. Those who were poor, corrupt, and incompetent were accused of causing society's disrepair. The faults must lie in their blood. Why should everyone be at the mercy of the genetically impaired?

It was former U.S. president Theodore Roosevelt who imparted to the public an ideology now eerily similar to the principles of Nazism. "The great problem of civilization is to secure a relative increase of the valuable as compared with the less valuable or noxious elements in the population," he said. "This problem cannot be met unless we give full consideration to the immense influence of heredity."[27]

In 1909, the American Breeders Association created its Committee on Eugenics. Its mandate: to move beyond the study of heredity, focusing specifically on race and ethnicity. A year later, Indiana became the first state to legislate forced sterilization of prisoners, the poor, and those with psychological problems. In 1909, Washington allowed the forced sterilization of "habitual criminals" and rapists for the "prevention of procreation." California was the third state to enforce involuntary sterilization.[28]

Yet in 1910, one of Sir Francis Galton's star students, mathematician Karl Pearson, confessed the weakness of the eugenic scientific argument. Concerning a paper on alcoholism and its relation to eugenics, Pearson stated, "The writers of this paper are fully conscious of the slenderness of their data. . . . They will no doubt be upbraided with publishing anything at all, either on the ground that what they are dealing with is 'crude and worthless material' or that as 'mathematical outsiders,' they are incapable of dealing with a medico-social problem." In a footnote, Pearson further questioned this new field of science when he quoted a critic saying, "The educated man and the scientist is as prone as any other to become the victim . . . of his prejudices. . . . He will in defense thereof make shipwreck of both the facts of science and the methods of science . . . by perpetrating every form of fallacy, inaccuracy and distortion."[29]

But for years, that was all it was: a footnote. By 1917, 17 states had legislated forced sterilization to stop the "dregs of society" from passing along their "illnesses." In the United States, from 1907 to 1921, 1,853 men and 1,380 women were sterilized. Approximately 2,700 insane patients, 400 feeble-minded and 130 criminals were operated on.[30] During the Great Depression, hospital budgets were slashed, and staff decided to sterilize young women with mild psychological problems. Between 1929 and 1935, 14,651 Americans were sterilized—9,327 women and 5,324 men.[31]

Eugenic practices were not confined to the United States. During the 1930s, Canada, Germany, France, and Japan were among the countries to create their own sterilization laws. In Canada, between 1928 and 1972, 2,900 forced sterilizations were ordered under the province of Alberta's Sexual Sterilization Act. Nellie McClung, who was instrumental in winning Canadian women's right to vote, praised the Alberta government for its "foresight and courage" in introducing the act.[32] Since then, the act has been revoked and Alberta has paid more than $800 million in compensation to thousands of individuals. British Columbia's sterilization law was in effect from 1933 to 1973.[33]

Yet it was Germany that brought new meaning to the concept of eugenics. According to the Central Association of Sterilized People in West Germany, from 1934 to 1945 the Nazis sterilized 3,500,000 men, women, and children.[34] Adolf Hitler leaned on eugenics in order to justify his loathing of Jews. He believed that they would dirty his thousand-year Reich, which was to be populated by the Aryan race. To convince the public of his ideologies, he transformed his personal hatred into a "scientific fact" that became dangerous to dispute. He read everything he could on American eugenics law, convincing German officials of its merit, and thus setting the stage for genocide.

When Hitler invaded Poland on September 1, 1939, the Reich was in dire need of hospital beds. Overnight, eugenics metamorphosed into euthanasia. Between 50,000 and 100,000 Germans were seized from psychiatric hospitals and retirement homes, and gassed.[35] Hitler stalked his prey across Europe, kidnapping, torturing, and killing Jews and other "undesirables," launching what would become the most heinous crime against humanity in all of history. Edwin Black writes, "The war against the weak had graduated from America's slogans, index cards and surgical blades to Nazi decrees, ghettos and gas chambers."[36]

The United States, praised by Hitler for introducing him to the ways of eugenics, was desperate to dissociate itself from the tyrannical Third Reich. Eugenics had become an affront to humanity, and after further examination it was proclaimed a scientific farce. The Eugenics Record Office revealed that its years of records were worthless to the field of genetics because they focused on traits such as character, sense of humour, and self-respect, which it concluded were immeasurable.[37]

After the Second World War, the American Eugenics Society was reborn under a new president, Frederick Osborn. Race and social status were no longer to be studied. Life could be understood by studying our genes. Genetics would tell us who we really were, and in the process eugenics would be thankfully forgotten. In the 1950s, medical schools launched genetics departments and "heredity counselling clinics."[38]

---

While genetics has over the years challenged society's ethics, it has also propelled a medical revolution. It was less than two decades ago that Lap-Chee Tsui and his team at Toronto's Hospital for Sick Children mapped the mutation for cystic fibrosis, the most common genetic disease among Caucasians, with one in 25 carrying the CF gene. In Canada, carrier testing is offered to those with a family history of the disease. The United States has the world's largest screening program for cystic fibrosis. Expecting mothers are offered testing for the disease, regardless of their level of risk. Even those whose families are free of the disease are advised to have the test, which is a first in genetics.[39]

Cystic fibrosis is a devastating disease that afflicts approximately 3,000 Canadians, 30,000 Americans, and 20,000 Europeans.[40] The disease attacks the pancreas, lungs, and sweat glands. During infancy, children suffer from chronic lung infections and have a high salt content in their sweat. Most die before they reach 30. Lap-Chee Tsui, now vice-chancellor of the University of Hong Kong, has dedicated his career to the study of cystic fibrosis.

And after many years of research, he, along with his colleagues, discovered the deadly genetic misprint.

Born in Shanghai in December 1950, Tsui grew up on the Kowloon side of Hong Kong, in the small village of Dai Goon Yu. The budding scientist was mesmerized by nature, spending countless hours exploring ponds, looking for tadpoles and fish on which to perform experiments. On his college entrance exam Tsui earned an A in biology, and he was accepted by the department at the Chinese University of Hong Kong. He remained at the university for many years, earning a bachelor's and master's degree.

Tsui moved to the United States to pursue his Ph.D. at the University of Pittsburgh. Trained as a molecular biologist, he yearned to apply science to improve everyday life. He discovered medical genetics, and focused on cystic fibrosis because of its strong genetic component. In 1981, Toronto's Hospital for Sick Children was taking different approaches to study the hereditary disease. So Tsui made his way there, where he would eventually be appointed geneticist-in-chief and would head its Genetics and Genomic Biology Program.

In the early 1980s, Tsui and others were still trying to unravel the mysteries of cystic fibrosis. The first step was to pinpoint which chromosome housed the gene. As in Nancy Wexler's research, they used genetic markers to randomly search the entire genome, looking for one that would lead them to the corresponding disease gene. The study focused on 50 families that had between two and five members affected by the disease. Again, critics said the process could take a lifetime to complete.

But in 1985, Tsui, along with Dr. Manuel Buchwald and other scientists, proved critics wrong. The marker linked to cystic fibrosis lay in a region of chromosome seven, which was approximately one-hundredth of the genome. The region was vast, and locating the CF gene was a daunting task. Even the exact nature of the disorder was still in doubt.

With slow, cautious steps, the scientists began to search chromosome seven for the disease gene. Imagine these scientists standing in a line at one end of a field. With every step forward, they move to investigate a new section of grass . Each set of hands sifts through the blades of grass, each gaze is intent on anything that looks out of the ordinary. Their critics were right. It could take a lifetime.

Tsui had devised a unique sleuthing method for his scientists, called chromosome walking, which would later become a famous gene mapping technique. Human DNA was broken into small and large pieces. These pieces were placed in individual bacterial cells, which share the bases A, C, G, and T with human DNA, and thus cannot distinguish between their own DNA and that of humans. Unaware that the transplanted DNA was foreign, the bacterial DNA replicated with these newfound bases.

The exciting result was millions of different bacterial cells, each with its own corresponding piece of human DNA. As, Ts, Cs, and Gs of human and bacterial DNA locked together to form whole pieces of DNA. This collection of DNA clones created a library that, in theory, contained all the DNA sequences from the original human source. Using radioactive tracers, the library of clones was screened for the target DNA pieces. Isolating overlapping pieces of DNA, Tsui "walked" from one piece to the next, making his way across the entire chromosome.

The work of Bob Williamson and Ray White helped narrow the search when they identified two markers that flanked the CF gene. Yet the hunt continued to move at a snail's pace. And even when they turned in a new direction, Tsui and his team had to cover a large distance to determine if they were in fact moving closer to their treasure. It is like relying on a compass to find your way, says Tsui. If you walk a few metres, this won't give you a different bearing. You have to walk some distance to see the compass shift. At this rate, Tsui wondered if the gene would ever come out of hiding.

In 1987, Tsui approached Francis Collins, renowned for his work in gene mapping, at the American Society of Human Genetics' annual meeting. This sparked a collaboration that would, two years later, result in the mapping of the CF gene. Collins understood the value of chromosome walking, but it was an arduous process that tested scientific patience. To save time, Collins proposed chromosome jumping, which enabled the geneticists to jump over long pieces of DNA. An extension of chromosome walking, jumping is a technical trick that circularizes the DNA, which allows scientists to focus only on the end pieces, jumping over the middle sections altogether.

Collins created a library of these genetic circles, and screened for successful jumps. For many exhausting months, Tsui sent pieces to Collins for screening. There were countless unsuccessful jumps as they made their way across the vast field of DNA. At the Hospital for Sick Children, scientists including Johanna Rommens and Batsheva Kerem spent endless days and nights looking for something that was missing in the DNA of CF patients. Another scientist, Richard Rozmahel, spent his days in Lap-Chee Tsui's lab, standing over a printer that spat out DNA variations to analyze—a frustrating process that seemed a test of his fortitude. Then, in the evening of May 9, 1989, Rozmahel glanced at a printout and saw something different. Something was missing. It was a three-base pair deletion. (DNA carries instructions to make protein using a three-piece message called a three-base pair. Rozmahel had found a missing, or deleted, three-base pair.) For five months, Tsui, Rozmahel, and the Hospital for Sick Children research team vigorously screened for the mutation in DNA from 100 CF patients and 100 individuals free of the disease. In September 1989, they announced the mapping of the CF gene. This was the first step toward finding a treatment to slow the fatal effects of the disease.

We know now that this mutation accounts for 70 percent of CF patients. The remaining 30 percent can have any one of 1,300 different mutations that, so far, continue to elude gene hunters like Lap-Chee Tsui.

---

Today, the science of genetics holds unbelievable power over humanity. With the prick of a needle and the quick draw of blood, the biology of an entire life can be plotted and mapped. As life is just beginning, we can predict how and when it is likely to end. Genetic testing sifts through a person's DNA to determine whether he or she carries deviant versions of genes or proteins. Abnormalities vary from the large—missing an entire chromosome—to the minute—a flip, or reversal, of one pair of bases. Genes can be overactive, inactive, or missing entirely.

Even people who will not develop a disease can benefit from carrier testing, which may show whether they harbour genetic mutations that put their children at risk. Although someone with only one abnormal copy of a gene for a recessive disease is simply a carrier, and will never become afflicted, two parents who are recessive carriers may both pass on the gene—and hence the disease—to their offspring. About 870 of the 6,000 known single-gene disorders can be predicted. Common ones include Tay-Sachs disease, which afflicts one in 960 Ashkenazi Jews; cystic fibrosis, which affects one in 2,500 Caucasians; and Huntington's disease, which strikes one in 10,000 to one in 20,000 of people over all.[41]

Another form of testing—predictive testing—typically determines the possibility, rather than the certainty, of contracting an illness. Susceptibility varies with every disease. Women carrying the BRCA1 gene, for example, could have an 80 percent chance of developing breast cancer, which affects one in nine women. People carrying a gene predisposing them to coronary artery disease, which affects one in four, may have only minimally increased risk, because of the crucial role lifestyle plays in triggering the onset of this disease; smoking, obesity, and high blood pressure can be as influential as a predisposing gene. Those who have one copy of the CCR5 genetic mutation have a better chance of fighting the AIDS virus, while those with two copies are immune.[42]

Prenatal testing is very popular with mothers who are concerned their babies will be born with either a chromosomal abnormality or an inherited disorder. During amniocentesis, which has been offered since the 1960s, a needle is guided into the womb to extract amniotic fluid and test for more than 150 disorders, including Down syndrome and spina bifida. Approximately 2 percent to 3 percent of children are born with a significant birth abnormality, with about 0.6 percent of these due to chromosomal changes, 1.4 percent attributed to single-gene disorders, and the remainder a result of combined genetic and environmental factors.[43] Women are typically the caretakers of their children's health, before and after birth. It is expected that they will do everything in their power to ensure the health of their babies. "With prenatal diagnosis presented as a 'way to avoid birth defects,' to refuse testing, or perceive no need for it, becomes more difficult than to proceed with it," says Abby Lippman, of the Department of Epidemiology and Biostatistics at McGill University.[44] Women who fall into the age category for risk will likely feel obliged to have a prenatal test, even if they have no history of severe genetic disease.

But the most widespread genetic testing is of newborns. One common example is screening for phenylketonuria (PKU), an enzyme deficiency that, if left untreated, can cause severe mental dysfunction.[45]

In both developed and developing countries around the world, genetic testing is routine. Yet international guidelines for the use of genetic information and testing practices are severely lacking. From country to country, different methods determine susceptibility to genetic disease. Reporting practices also vary, making it difficult to disseminate information across borders.

While many countries in the Organization for Economic Co-operation and Development (OECD) follow standard laboratory guidelines in other areas, such regulations are not being implemented for genetic testing. Not all OECD countries insist on the accreditation of genetic laboratories—"home-brew" tests are offered by clinical research labs and are lacking in standardization—and supervision by trained medical staff is inconsistent. Errors have become commonplace due to improper handling and interpretation of data, and could have irrevocable consequences for those being tested.[46] Inaccurate results could well convince a woman to terminate a healthy pregnancy. Moreover, it must be stressed that, for many genetic diseases, testing can only predict risk. A positive result may not be proof that the disease will develop, even though it is often viewed as just that. In many cases, diet, lifestyle, even geography, interact with genetic tendencies to trigger disease onset. And sometimes life is simply left to chance, with diseases developing unexpectedly.

It has been said that DNA is not a blueprint of a person; it is merely an outline of what he or she may later fill in. Lippman says that genetics attempts to reduce the human condition to its genetic determination, a process she calls geneticization: ". . . differences between individuals are reduced to their DNA codes, with most disorders, behaviors and physiological variations defined, at least in part, as genetic in origin.[47] Such genetic determinism has many concerned that a new form of eugenics is on the rise. This time, power is being wielded not by government, but by expecting parents. The growing fears focus on a new procedure called preimplantation genetic diagnosis (PGD), which aims to prevent genetic diseases, but also enables parents to delete or tweak unwanted traits in their future children.

The first PGD baby was born in London, England, in 1992, to Michelle and Paul O'Brien. The couple already had a son, who was diagnosed with cystic fibrosis when he was only two months old. They wanted another child but didn't want to bring another person into the world with the fatal disease. When they learned of the experimental genetic screening, they signed up, despite scathing criticism that ran in the papers. "If they don't know, if they don't have a CF child of their own, how can they say whether it's right or wrong?" the O'Briens asked. "It's one thing to choose hair or eye colour—it's another to choose to prevent suffering.[48]

The couple decided to go ahead with the PGD screening. Eggs were taken from Michelle's womb and were fertilized with her husband's sperm. The embryos began dividing, and after the eighth division scientists pricked each embryo and released a cell for testing. Thousands of copies of every cell were made, and each group of cells was tested for cystic fibrosis. Those with the genetic predisposition were rejected. Two healthy embryos—one with two normal genes and the other with one normal and one CF gene (CF is recessive, so you need two copies of the gene to cause the disease)—were implanted. One embryo developed normally, and a girl—Chloe—was born free of cystic fibrosis.

Today, about 50 fertility clinics offer PGD, with two-thirds located in the United States. As many as 10,000 babies have been born after PGD screening. Disability rights groups are denouncing such selective breeding, saying that choosing one embryo over another is discriminatory and depreciates the lives of the disabled. How are those with Down syndrome to feel when they discover that countless parents would rather abort than have a child like them? Why are some diseases—and therefore people—easier to assimilate into "normal" society than others? "Women are expected to—pressured to—abort pregnancies when fetal disability is diagnosed," says Tanis Doe, a professor of social work at the University of Alberta. Doe, who is deaf and confined to a wheelchair, says approximately 89 percent of Canadian parents whose child is diagnosed with Down syndrome choose to abort.[49] According to the International Sub-Committee of the British Council of Disabled People, there are more environmental causes (such as traffic and work accidents) than genetic causes that result in disability.[50] It should also be pointed out that cells do not express all of their genetic characteristics. A child might "grow out" of a disorder before birth, or may experience minimal symptoms.

Since the 1970s, however, many women have exercised the right to abort unwanted pregnancies. Should this right be overshadowed by the rights of an embryo predisposed to a disability? A distinction must be made between those with a disability and the disability itself. As noted by C. Cameron and R. Williamson in the *Journal of Medical Ethics,* many parents who choose to abort "do not see themselves as making a moral judgement about the worth or rights of people living with that genetic condition."[51]

But what if parents decide to go ahead with the pregnancy, and the afflicted child decides life is not worth living? As science continues to outpace the law, genetics has created a legal conundrum: the wrongful life suit. A contentious court decision in France in 2001 ruled that children with Down syndrome have a legal right never to have been born, and that physicians are liable if they do not diagnose the disease during screening. Ethicists, lawyers, and physicians were outraged, including Margaret A. Somerville, founding director of the McGill Centre for Medicine, Ethics and Law: "It's a better-off-dead ruling. The law has always upheld that life is better than no life and now the court has concluded the opposite. To award damages, you must think that the harm of being born outweighs the benefits."[52]

The case centred around Lionel, born with Down syndrome in 1995, and his mother, who said she would have aborted if she had known the risk to her unborn child. A panel of three judges decreed that children should be compensated for a disabled life if their mothers were unable to abort once prenatal testing revealed a genetic risk of a disability. Since Lionel's mother's right to have an abortion was hindered, the judges ruled that her doctor was responsible for the cost of Lionel's care. Dr. Roger Bessis—president of the French College of Echography, which regulates the country's ultrasounds—says that in 2000, 90 percent of genetic abnormalities were detected in prenatal screening in Paris, and 8 percent of these mothers did not abort. "The courts said the doctor was 100 percent liable," says Bessis, "but everyone knows that medicine can never be 100 percent accurate. We do the best we can."[53] In 2000, the French court ruled that a disabled boy named Nicolas be awarded damages from his mother's physicians, who had failed to mention that her baby was at risk because she had contracted German measles. The result: Nicolas was born with a weak heart, deaf, and unable to talk.

In some cases, genetic testing infringes on the autonomy of a future individual even when abortion is not involved. Autonomy encompasses liberty, privacy, and individual choice. Above all, it is the "personal governance of the self that is free from external control." And although "children may not possess the legal capacity to assert their autonomy, they nevertheless possess the 'right' to exercise autonomy in the future." So while children are unable to make autonomous decisions during the beginning of their lives, their future right to do so always exists.[54] Genetic testing, therefore, should be done for the welfare of the child, rather than the interests of the parents. Just because a mother and father want to know if their child has the genetic defect for, say, Huntington's disease, this does not mean their child will want such foresight into the future. The child could argue that the parents have violated his or her right to genetic ignorance, and that his or her life is forever changed because of the imposed decision. Despite the availability of testing for Huntington's, only 10 percent to 15 percent of at-risk adults choose to know their fate. Yet while the majority of adults opt not to read their genetic horoscope, the choice to do so has been made for countless children.[55]

Much prenatal and postnatal testing is done, however, for the direct health benefits of the child. Many disorders, including PKU, are treatable through early detection. The moral and legal dilemmas lie with non-therapeutic testing. While PGD could in theory erase genetic disease from society, it also enables parents to design their own babies. In the future, embryo screening will inevitably become cheaper and widely available. With the human genome now mapped, the possibilities for testing are infinite. One day, parents could present their physicians with a shopping list of preferred traits. Those embryos that didn't make the genetic cut could then be discarded.

Should parents be allowed to engineer their gender of choice? In some countries like India, being a woman is a hindrance. But is it a genetic disorder worthy of deletion? In terms of health, females tend to live longer, and men are not biologically superior. Yet in Indian society women are at a social disadvantage. Gender discrimination is common, and sex selection has tipped the sex ratio of the Indian population.[56] The fear with PGD is that it will be used for sex selection in both the developing and developed world.[57] Again, the autonomy of the child is challenged.

According to the Supreme Court of Canada, "[a]ny right or interest the fetus may have remains inchoate and incomplete until the birth of the child."[58] It could be said, then, that parents may choose prenatal rather than postnatal testing without violating the rights of their child. Genetic manipulation to create "designer babies" is done at the prenatal stage, and thus doesn't require the permission of the future individual. Could it be said that the Canadian Charter of Rights and Freedoms' "liberty and security of the person's interests" allow mothers to manipulate and use their bodies as they please, even altering the genetic identity of their fetuses? It has been concluded that genetically altering a life for non-therapeutic or cosmetic purposes is not regarded as a fundamental right that deserves the protection of the Charter.[59]

Many parents argue that they are being selfless when they genetically stack the cards in their children's favour. The desire to have a baby unencumbered by pain and disease goes beyond self-interest. Competition for work, love, even a house, is fierce, and bad genes can leave a person destitute. Rabbi Joseph Ekstein is a parent who intimately understands the calamitous power of bad genes. In Brooklyn's Crown Heights community, where Ekstein lives, one in 16 people is a carrier of a single copy of the Tay-Sachs gene, compared to one in 300 in the general U.S. population.[60] If two carriers marry, each of their children will have a 50 percent chance of being a carrier, and a further 25 percent chance of dying of the disease. Four of Ekstein's own children have died of the disease. When the Tay-Sachs gene was discovered, young people who carried the gene found their future happiness threatened by their secret. If their status was known, who would ever marry them? What would become of the Crown Heights "marriage market"? Ekstein had a plan.

While the rabbi understood the necessity of testing young people for the Tay-Sachs gene, he also realized that, if he could prevent any two carriers marrying, in theory the community could eventually be freed from the disease. He proposed that genetic testing be anonymous. If two carriers intended to marry, Rabbi Ekstein pronounced the match unsuitable—which could be for a variety of reasons—and under parental guidance, the match was dissolved. The carriers eventually found non-carriers to marry, and the disease remained latent. In 1983, Ekstein launched the Association for an Upright Generation, which travels to high schools to test teenagers long before they are considering marriage. Each teen is given a number that coincides with his or her results, which remain anonymous to everyone but the association. A computer in New York houses the results. When someone is considering a match, one phone call to the data centre reveals the couple's "marriageability." Tens of thousands of Jews have been tested for Tay-Sachs, and are now being screened for cystic fibrosis.

Genetic testing is crucial to medical progress and individual health. A simple blood test reveals someone's susceptibility to a disease, and the person's lifestyle can be adjusted accordingly. While general medical records disclose a great deal about a person's present health, genetic data goes one step further to predict not only the problems of today but also those of the future. While our genes may be influenced differently by lifestyle factors at certain times in our lives, the genes we have inherited dictate our predisposition to disease.

Yet a faulty gene is not always a bad thing to inherit. Take, for example, the mutation for the gene HbS. Two copies of it will cause a person to develop sickle-cell anemia, a debilitating blood disease that results in joint pain and clotting. Yet one copy acts as a genetic shield for malaria, a fatal disease that is transmitted through mosquito bites. If the mutation predisposing to sickle cell anemia is erased, so is our defence against this other debilitating condition. Rearranging our genes is a complex issue that doesn't offer easy answers. There are just more options to weigh.

---

While physicians and genetic counsellors use genetic information in the name of science, however, insurers and employers have a much different agenda. When hiring new workers, it may be in an employer's interest to learn their genetic predisposition to serious—and expensive—diseases. How much will this individual cost the company? Will he or she be a drain on the benefits plan?

But there is a danger that false conclusions may be drawn from a genetic reading. Who is to be responsible for evaluating DNA and predicting the future? Does this person have an understanding of genetics, its certainties and ambiguities? Someone who carries the BRCA1 susceptibility gene, for example, may never develop breast cancer. But will an employer or insurer care that this is a mere possibility, or simply reject the potential financial burden? In the United States alone, according to the OECD, 183,000 women develop breast cancer annually, with 41,000 dying from the disease. The cost: more than US$10 billion.[61]

In the past, some employers have requested genetic testing to protect their employees from specific job-related risks. It is important, for example, to be aware of a worker's "hypersusceptibility" to dust or chemicals found in the workplace. A certain type of contaminant could have very diverse effects on different individuals. Some people may seem immune to a given chemical, while others develop cancer.

An excerpt from a report by the U.K.'s Nuffield Council on Bioethics reveals that as far back as 1938, companies were exploring the impact of heredity on working conditions. "The majority of potters do not die of bronchitis. It is quite possible that if we really understood the causation of this disease we should find out that only a fraction of potters are of a constitution which renders them liable to it. If so, we could eliminate potters' bronchitis by regulating entrants into the potters' industry who are congenitally exposed to it."[62] Employers have long sought ways to reduce costs and heighten productivity. Healthy employees are far cheaper than unhealthy ones. They arrive on schedule; substitutes are rarely needed; costly workplace precautions are avoided; and they focus on their jobs rather than on personal health problems. Not to mention the money at stake: 5 percent of health-care claimants exhaust 50 percent of all medical resources in any jurisdiction, and 10 percent of claimants drain 70 percent of all resources.[63]

While screening for personal susceptibility may reduce heath costs, however, it's far from a perfect situation. Regardless of their susceptibility, all those who are exposed to toxic agents are at risk of becoming ill. While genetic testing may protect the few, others may suffer from lax precautions.[64]

There is a distinct conflict between an individual's right to privacy and an employer's desire to create an efficient, cost-effective, and profitable business. In essence, privacy is "control over when and by whom the various parts of us can be sensed by others."[65] The use of genetic test results without the individual's knowledge is an invasion of privacy. The person's genetic identity is revealed without consent, infringing on the person's autonomy. The right to privacy safeguards our basic rights to not be looked at, listened to, harmed, hurt, or tortured. Privacy also prevents prejudice and stigmatization in society, and preserves "degrees of intimacy" in friendship and love.[66]

To uphold their individual privacy, people must be able to exert their right to choice, secrecy, and confidentiality. The ability to choose allows workers to refuse genetic testing without the risk of being punished or losing their jobs. Secrecy ensures privacy once a test has been performed;

the results remain undisclosed for as long as individuals wish. The right to confidentiality protects people from secondary disclosure of genetic results. The Hippocratic oath, which is taken by all physicians, prohibits doctors from disclosing information about their patients.[67] While neither the Canadian nor the American Constitution affirms a right to privacy, the supreme courts of both countries deem privacy to be deserving of constitutional protection. In Canada and the United States, information privacy is protected by the forbiddance of "unreasonable searches and seizures."[68]

The Supreme Court of Canada safeguards the privacy of personal information from third parties, with the intention of protecting a "'biographical core of personal information' by granting to the individual control over the dissemination of this information at the point of disclosure."[69] Canada's Personal Information Protection and Electronic Documents Act (PIPEDA)[70] controls the collection, use, and disclosure of personal data by private corporations. Under PIPEDA, companies must have a person's informed consent to collect, use, or disclose such information.[71]

In Canada, citizens rely on a universal health care system. This is not the case in the United States, where employers cover the majority of health care costs. Of approximately 250 million people in the U.S., 150 million have private group insurance as an employee or employee's spouse or dependent. About 10 to 15 million have their own policies; 33 million, mainly elderly, are covered by Medicare, the federal health insurance program for those aged 65 or older and those who are disabled, while 23 million of the poor rely on Medicaid, the national health insurance program for low-income families. This leaves 34 million without any medical insurance at all.[72]

Whenever people have to seek their own insurance, the new information available from genetics adds a serious complication. Private insurance companies sell financial coverage for life's unforeseen medical events, which include disability, long-term care, and death. Insurance is critical in protecting a person from the costs of serious injury. Yet with respect to life insurance, it is not only a matter of life and death. Ineligibility for insurance can disqualify someone for a mortgage and wipe out job prospects. And while people are doing everything in their power to secure insurance to protect themselves and their children, insurers are intent on exposing their medical weaknesses, and predicting the cost of their future claims. A potential insuree is slotted into the category of good risk, where the chance of claim is slim, or bad risk, where there is a high chance of claim. To anticipate risk, the insurer assesses the probability of illness or early death, based on the individual's age and occupation, as well as lifestyle choices. Does he exercise? Is she obese? Medical tests may be ordered before the applicant's health status is determined. The answers determine the cost of the individual premium. If the person suffers from a serious illness, he or she may be denied coverage, or the plan may exclude the costs of a certain disease.

Sometimes an insurer knows considerably less about the person's health than he or she knows. In the past, some applicants have not revealed smoking or excessive drinking, which can lead to serious complications or even death. Genetics has intensified the discrepancy between what the insurer can foresee and what the applicant may know. Through genetic testing, someone may learn that he or she is predisposed to a late-onset disease like Parkinson's, and may maximize an insurance policy to help cover the anticipated cost of treatment. Paying out countless high claims could destroy the insurance industry—especially since people with little apparent genetic risk might save money by forgoing insurance. Premiums could become so high that even risky recipients might take their chances with future health care costs, banking their premium money and hoping to cover the costs themselves.[73]

Genetic disclosure to employers and insurers poses another dangerous risk: infringing on the individual's right to privacy. Guidelines for the use of genetic information have been passed by the European Council and UNESCO, while in the U.S. over half of the states have enacted restrictions on the use of genetic data by insurers. The American Health Insurance Portability and Accountability Act of 1996[74] forbids group insurers to apply "pre-existing condition exclusions" to genetic disorders that are asymptomatic. In 2000, former president Bill Clinton signed an order that forbids U.S. federal agencies from using genetic data as a consideration for employment or promotion.[75]

In the United States, the Americans with Disabilities Act of 1990[76] protects citizens from genetic discrimination. According to this legislation, a disabled person has "(a) a physical or mental impairment that substantially limits one or more of the major life activities of such individual; (b) a record of such an impairment; or (c) being regarded as having such an impairment." It has been said that those predisposed to a genetic disease, who may not presently show symptoms of it, fall into the latter category.[77] American employers have avoided prosecution under the act by proving that a disability prevents an employee from fulfilling a job function, and showing that accommodating disabled people would create an "undue hardship" for the company—which means that the person is unable to fulfill the qualifications for employment. Employers may also show that their decision to terminate or curb someone's employment was based on another reason altogether.[78] It must be proven, in that case, that the decision was based on genetic data rather than on disclosed general medical information.

Even if companies can avoid the provisions of the Americans with Disabilities Act, they must abide by the rules of informed consent. For genetic testing to occur, an individual must provide free and informed consent to be screened for a certain condition, according to the World Health Organization (WHO). This includes genetic counselling beforehand that explains what they are being tested for, which should be followed by counselling once the results have been disclosed.[79]

Informed consent took centre stage in the case of *Equal Employment Opportunity Commission (EEOC) v Burlington Northern Santa Fe Railroad Co.*[80] In 2001, 125 of the railroad's 40,000 employees filed disability claims for work-related carpal tunnel syndrome, which may be caused or aggravated by repetitive hand or wrist motions. Hammering spikes and working railway machinery was thought to be the culprit.[81] Burlington's medical director read about a genetic susceptibility for carpal tunnel syndrome. Without their knowledge or consent, many claimants were given a genetic test to determine their predisposition to the disorder, in the hope that Burlington could refute the claim that the injuries were work-related. (Workers were tested for a deletion on chromosome 17, which—according to Dr. Phillip Chance, the creator of the genetic test—was a nonsensical choice. The chromosome 17 test was designed for a condition called hereditary neuropathy with liability to pressure palsies (HNPP), which he says rarely results in carpal tunnel syndrome.[82])

At least one worker who suspected genetic testing and refused to give a sample of his blood was threatened with termination of employment. The railroad's secretive practices were revealed when another worker's wife, a nurse, inquired as to why several vials of blood had been taken from her husband during a mandatory medical exam.[83] The EEOC lashed out: "The commission takes the position that basing employment decisions on genetic testing violates the ADA. . . . Any test which purports to predict future disabilities, whether or not it is accurate, is unlikely to be relevant to the employee's present ability to perform his or her job."[84] In February of 2001, the EEOC settled its first court action challenging workplace genetic testing under the ADA. Burlington paid US$2.2 million in damages to employees,[85] ceased its testing program, and acknowledged the need for national legislation on genetic testing by employers, agreeing to send written statements of this opinion to Congress and the president.[86]

The world was finally catching up with science. Politicians and lawmakers were addressing the social and legal concerns posed by genetics. Citizens were beginning to grasp the technology, and how it could affect their lives. But not everyone was pausing for reflection and self-congratulation. A brash, outspoken Icelander named Kári Stefánsson refused to keep to the measured pace of North American progress. He left the United States to return home with a vision. He would transform his small island into a genetics laboratory, and decode life's remaining biological riddles. He would lay bare the scientific mysteries of humanity for all to see, and his fellow Icelanders would christen him a hero. There was a new code to crack.

## Chapter Three

# A New Code to Crack

## Population Genetics and the Genome's Instruction Manual

There is an island anchored in the North Atlantic, on the edge of the Arctic Circle. It is a place of myth and legend that belongs in children's fables. Out here in the ocean, the sky is as vast and surreal as in a picture book, reaching as far as the imagination, eventually dipping into the glassy water. In the midst of this crystalline blue is a small shadow of land—as if a piece of the moon has dropped from the sky. Here and there in the moonscape of craggy black lava, towering volcanoes lift the brittle ground. A series of twisting rivers trail down the sides of the mountains. Cutting into the horizon are jagged glaciers, their icy peaks reflecting the blue sky. This small island is pounded by rain and rattled by earthquakes, but over the past thousand years it has endured much more. This is a land of plague and famine, of heroes and villains, rich in Viking history, and now famous for its genes. This is Iceland.

Over the past few years, Iceland has been placed under the microscopes of scientists who are fascinated by its homogeneous population. Almost all of the country's 275,000 residents are descendants of the first settlers, who arrived here in the year 874. Most of the settlers came from Norway, with a small percentage making the journey from Scotland and Ireland. Old Norse, the language of Scandinavia, was adopted and later evolved into modern Icelandic, which is still spoken today.

Between the thirteenth and nineteenth centuries, Icelanders battled for survival. The climate was dreadfully cold, and the land was barren and infertile and impossible to cultivate. In the 15th century, 70,000 inhabitants succumbed to the plague. Three centuries later, smallpox killed thousands more. By 1800 there were only 45,000 people left, living without wood to burn or build houses, finding shelter in turf huts, and sleeping on mud floors.[1] Few foreigners visited, and few Icelanders left their secluded home. Such isolation preserved their language, their cultural traditions, and their gene pool.

Left with little more than their bare existence, Icelanders found solace in the stories of their ancestors. The first Icelandic genealogical record, dating back to 1125, still survives today. There is the religious history of the Icelanders, who converted to Christianity at the end of the tenth century, and there are biographies of celebrated bishops. There are the regal Kings' Sagas, which recount the lives of kings and queens, poets and adventurers. *The Legendary Sagas* are stories of myth and fantasy that hail gods and deplore demons. Influenced by French Romantic literature, these sagas have inspired centuries of Icelandic poets. The common theme among Icelandic legends is the justice of the individual defying the unjust world.

It wasn't simply the stories that fascinated Icelanders; it was the process of recounting them. They moved beyond the grand battles and triumphs of Odin, the "restless seeker of wisdom," or Thor, "he with hammer," and became engrossed in their own, more subdued past. Yet while Odin theorizes, "He should get up early, the man who means to take/another's life or property;/the slumbering wolf does not get the ham,/nor a sleeping man victory," a typical Icelander's story is more clinical in its message.[2] The genealogical record for a man born in 1740 reads, "Half of his body had contracted in the autumn, and so had his tongue. As a result, his speech was barely recognizable." The story of a man born in 1755 states, "Among his descendants, certain family characteristics have persisted, including tremendous energy, endurance and dexterity."[3]

Icelanders recorded their own histories because they yearned for a sense of identity, one that would not be wiped out with the next famine or drought. They created intricate family trees dating back to the first settlers. Their lineages were mapped, along with their health and social status and, eventually, their cause of death. One day they would be gone, but their stories would live on.

Modern Vikings carry briefcases instead of broadswords. The capital city of Reykjavik has been invaded by sports utility vehicles, cellphones, and McDonald's. But the people remain passionate about their history. Handwritten records have been translated into computer code, and the history of the country is captured on one massive, downloadable digital file. In today's genomic era, where scientists must often use the DNA of random individuals to find disease genes, these family histories have become a multi-million-dollar commodity. And one Icelander is poised to cash in on his country's rich past.

---

Brandishing plans for the world's most extensive population database, Dr. Kári Stefánsson returned to his native land like the hero in one of those Icelandic sagas.[4] Like most Icelanders, he had left his country in search of fame and fortune. Iceland has only one university, and science positions are in short supply. But in the mid-1990s Stefánsson came back, bringing with him millions in American venture capital and a business plan that would catapult Iceland into the big leagues of the biotechnology world. He promised jobs, money, even celebrity to this isolated community. And despite a global outcry from ethicists, lawyers, and fellow scientists, he has kept his lofty promises, while delivering much more than his fellow Icelanders had bargained for.

Stefánsson is striking, with eyes as blue as the North Atlantic and sprigs of once-flaxen hair now matured to a dignified grey. He avoids tobacco, opting for gum or Diet Coke, and spends hours at the gym, hoisting barbells and running around a basketball court.[5] Although some consider him hot-tempered and unforgiving, Stefánsson is also a hopeless romantic, with scrolls of memorized verse tucked away in his neurologist brain.

His father, Stefan Stefánsson, was a writer and socialist politician whom his son calls a "social romantic who had a keen sense of history." When asked how his father would react to his controversial business venture, Stefánsson says, "I'm convinced that he would look at what we are doing as an attempt to seek poetic justice for the hardships our nation has suffered for the past thousand years."[7] Stefánsson is haunted by his father's premature death at 67, and his mother's at 62. His brother was tormented by the hallucinations and violent behaviour of schizophrenia. Rather than follow his father's romantic notion of becoming a writer, the lover of poetry found inspiration in science.

Stefánsson remained in Iceland for his medical degree, but did his specialty training in neurology and neuropathology at the University of Chicago. As a medical resident he examined the bodies of patients with amyotrophic lateral sclerosis (ALS), which attacks nerve cells (neurons) in the brain and spinal cord, resulting in muscle weakness, paralysis, and eventually death. His job was to find signs that hinted at the causes of the savage disease—often called Lou Gehrig's disease, after one of its victims, the American baseball legend. Between 1983 and 1993, Stefánsson was on faculty at the University of Chicago.

Then his specialty shifted to multiple sclerosis (MS), a neurological disease that causes tremors and paralysis. Stefánsson and a team of researchers analyzed proteins in diseased brains, searching for clues to demystify the condition. But little could be concluded from the proteins alone. Abnormal proteins are the result of disease genes; a blip in DNA churns out faulty proteins. Stefánsson's isolated proteins were not the cause of MS, but rather a bodily reaction to the disease.

In 1993, Stefánsson moved to Boston to take a post as a neurology professor at Harvard University. He continued his research on MS, but changed his approach. He would begin with the genetic error and work forward. His plan was to study the DNA of related individuals to see what genetic errors were passed from one generation to the next. Lap-Chee Tsui and Francis Collins had mapped the genetic mutation for cystic fibrosis (see Chapter Two), but CF is a single-gene disorder, and MS is a complex genetic disease that involves a combination of genes and lifestyle factors. If there was any hope of finding the MS gene, Stefánsson would need large families to study. Immediately he thought of his remote island homeland, with its isolated, homogeneous population. He would transform the obscure little country into a genetic laboratory, and bring biotech home.

In 1994, Stefánsson and Jeffrey Gulcher—then a University of Chicago grad student—made the trek to Iceland to take blood from neurologist Dr. John Benedickz's MS patients. They returned with the samples to Boston, where they were analyzed. Back in Iceland, Benedickz painstakingly compiled family lineages of MS patients.[8] The detailed records would reveal not only individual health, but also the environmental factors that greatly influence complex diseases like MS. Iceland's vast genealogical records could be used to uncover the genetic mystery of the disease.

While Stefánsson was away, another company began harvesting Iceland's precious history. In the early 1990s, FRISK Software International started the construction of an electronic genealogical database, tapping into the censuses of 1703, 1801, and 1910 as well as present government records. FRISK director Fridrik Skúlason, an internationally renowned computer virus hunter, shares the Icelandic passion for genealogy, and put his programming skills to the test. Skúlason has compared the process to "working out a puzzle the size of a football stadium, with half of the pieces missing and the rest randomly scattered."[9] Stefánsson partnered with Skúlason, offering more information to compile, and speeding up the arduous process.

The *Book of Icelanders* is made up of 12 censuses from 1703 to 1930. The database, however, only holds records of approximately 650,000 people, less than half the number of Icelanders born in the country since the first settlement. When Stefánsson's company uses the data, names are not attached to records; individuals are identified only by number. Records are also protected from hackers by an intricate encryption system. What makes this database so unusual and priceless is its "connectivity index"—the percentage of genealogical links between family members.[10] While Icelanders are fascinated by their connections to the past, it is Stefánsson who understands how precious these records really are. Not only could they offer clues to one day cure disease; Iceland could even profit from its people's genes.

In 1996, Stefánsson returned to Iceland to meet with David Oddsson, his old classmate and Iceland's prime minister. Stefánsson undoubtedly pointed to the country's unique historical records and its homogeneous population, and offered his proposal: link genealogy with medicine to create a population genetics company. Icelanders would no longer venture into the great big world looking for opportunity; it would be in their own backyard. The global science community would tip its hat to the progressive nation. And pharmaceutical giants would invest millions in a firm destined to change the treatment of disease. Iceland would become a bustling hub of biotech, shedding its image as a fishing port. Leave the cod be; there were precious genes to trade. By the end of the meeting, Oddsson was on board.

Stefánsson raised US$12 million in venture capital within a year and, along with his former student Jeffrey Gulcher, launched DeCode Genetics. Even though the start-up was funded by American money, Stefánsson insisted the company be based in Reykjavik. He rented the second floor of a Kodak processing plant, hired just over a dozen staff, and bought a handful of DNA-decoding and DNA-copying machines worth hundreds of thousands of dollars. It was time to get to work.

The neurologist convinced other physicians that the project wasn't a sellout, that they were not simply trying to make a quick buck off patients. It had the potential to advance Icelandic medicine. Just think how patient care would improve if doctors could home in on the genes that cause disease. Other neurologists, from Reykjavik Hospital and the National Hospital of Iceland, joined Stefánsson's team. They created family trees by talking to MS patients about their families' medical history, inquiring whether anyone else showed signs of the disease.

The neurologists mapped the lineage of 75 MS patients and 42 healthy relatives in 16 families. With their consent, individuals gave blood, which was sent back to the DeCode facilities. The DNA was analyzed using genetic markers similar to what Tsui and Collins had relied on to map the CF gene. One day the DeCode team was walking along chromosome three, searching for identical markers within MS families, when they found the genetic misprint. They had narrowed the search for the MS gene.[11] "When we find a locus, it's as if we know the gene is in Scandinavia. Then we will narrow the region to, say, a small country like Iceland," says DeCode scientist Thorgeir Thorgeirsson. "Next we can start going door-to-door searching for the person—or the mutation responsible for the disease."[12]

To home in on the MS gene, and for large-scale gene mapping in the future, Stefánsson needed more samples, and he intended to connect genetic data with patients' medical conditions. DeCode would link a genetics database, a genealogical database, and the nation's individual medical records, which date back to 1915. But Stefánsson was hesitant to embark on such a costly project without some legal guarantee that access would not one day be revoked. So he turned to the government of Iceland. As a result, a bill was drafted that would grant DeCode an exclusive 12-year licence to profit from the Health Sector Database (HSD). It was presented to the Icelandic Parliament in March 1998 as a *fait accompli,* but was immediately contested by many MPs.[13] How could DeCode ensure that such sensitive information would remain anonymous? Who would be given access to the medical data? And why should only one company be able to profit from Icelanders' common genetic wealth? Should this not be the property of the citizens themselves?

The most contentious aspect of the project was its consent process. If patients did not voice their disapproval, their personal medical information would be included automatically. Rather than opt in, they would need to opt out. Physicians shared the politicians' outrage. Patients must consent to the release of their own medical records, said the doctors, adding that if they released information without permission they would be breaching their Hippocratic oath, which promises, "Whatever, in connection with my professional practice, or not in connection with it, I see or hear, in the life of men, which ought not to be spoken of abroad, I will not divulge, as reckoning that all such should be kept secret."[14]

Many Icelandic physicians were furious that their personal files could be taken without their consent. One voiced his outrage: "I think it is a gross insult to my practice as a physician to appropriate documents that I record as memos and then to turn them into scientific data. This would be an assault if done by force. . . ."[15] And if doctors know that their patients can quickly gain access to personal medical notes, rather than going through the present arduous and expensive request process, these recordings might become less forthright and exact, argues Henry Greely, an opponent of the Icelandic project and a professor at Stanford Law School. Greely adds, "The failure of a doctor to note her suspicions of a patient's alcoholism in the medical record could later have dramatic, even fatal, consequences for the patient at the hands of another doctor."[16]

But—as with everything in life—the bad had to be weighed against the good. When Icelanders were wheeled into an emergency room with life-threatening injuries, perhaps unable to speak, physicians would be better able to perform their duties if they could refer to thorough medical histories. And Greely acknowledged that, as for long-term hospital stays, a team of health care professionals—physicians, nurses, anesthesiologists and often specialists—require patient records to do their jobs.

Other critics of the Iceland project lambasted Stefánsson for including children and mentally incapacitated individuals in the database unless parents or guardians opted out on their behalf. And under no circumstances were families able to opt out for deceased relatives.[17] Their information was included automatically because, it was argued, using it could do no harm to the dead.

After plans of the database were made public, Stefánsson had to run frantically around performing damage control. He made speeches to groups of Icelandic citizens whose opinions could affect the future of his business. Genetic testing required consent, he explained; it was just the collecting of medical records that was assumed to be permitted unless the person specifically refused. For centuries, he told audiences, researchers had relied on implied consent to use personal health information for statistical trends, because it was impractical to track down thousands of individuals.[18] According to the WHO, Iceland had the right and duty to collect medical data to survey communicable diseases and drug safety. Furthermore, he argued, any nation that offered universal health care (like Iceland) had the right to obtain medical records with the aims of monitoring and improving medical access, services, and utilization.[19] Miraculously, Stefánsson won over his fellow citizens.

This is not so hard to believe, says Halla Thorsteinsdottir, a professor of philosophy and a science policy researcher from Iceland now working at the University of Toronto's Joint Centre for Bioethics. In Iceland, health information is not considered highly sensitive—as it is in the United States—because the state health care system offers medical attention to all. And even if it is discovered that Icelanders are more predisposed to certain diseases, discrimination is not the country's main concern. Iceland is a resource-poor nation whose economy is largely based on fishing. Thorsteinsdottir agrees with a renowned genealogist who says that, metaphorically speaking, Icelanders find the gold but the jewellery-making is done elsewhere. It was time, now, to keep the gold at home.

Before the bill was passed in 1998, a Gallup poll concluded that 58 percent of Icelanders were in support, with 19 percent against and 22 percent undecided. In April 2000, 81 percent were in support, with only 9 percent opposed and 10 percent undecided. A mere 7 percent have chosen to opt out of releasing their medical records.[20] Icelanders' most pressing concern today, says Thorsteinsdottir, is whether DeCode can compete with other global leaders in biotech. If DeCode were to one day declare bankruptcy, hundreds of jobs would be lost and the country's economy would plummet.

In December 1998, the Icelandic Parliament passed the contentious Health Sector Database Act, granting DeCode its licence. Under the terms of the agreement, DeCode will "fully fund development and implementation of: (1) a system of computerized medical records in clinics and hospitals located throughout Iceland; and (2) the centralized collection of data and operation of the HSD. DeCode will pay all related expenses incurred by the government, plus an annual inflation-indexed payment of 70 million Kronur (currently about US$900,000 or US$3/Icelandic citizen), plus 6 percent of its annual pre-tax profits up to an additional 70 million Kronur. . ." This adds up to less than 0.5 percent of Iceland's public health expenses, which were 51 billion kronur in 2000.[21]

Even before the bill had been passed, DeCode was diversifying its business plan. Stefánsson was confident that DeCode would map many disease genes in the future, and wanted to cash in on these discoveries. Klaus

Lindpaintner, of the big pharmaceutical firm F. Hoffman–La Roche, had the same idea. After reading about DeCode's business plan in the *Wall Street Journal Europe,* Lindpaintner called a meeting with Stefánsson. The end result: a five-year, US$200-million contract giving Roche world rights to develop all drug and diagnostic products from DeCode's research. Roche was given a 10 percent equity stake in DeCode, and to win the support of Icelanders, the pharmaceutical giant promised that any drugs developed would be offered to them free of charge. DeCode bought the entire building that housed its one-floor operation, along with new equipment, and hired 250 people, employing almost one out of every 1,000 Icelanders.[22]

In July 2000, DeCode raised US$173 million from its initial public offering, or US$610 per Icelander. The stock (Nasdaq/Nasdaq Europe: DCGN) first sold for about US$25, but was hammered when the biotech bubble burst. In September 2002, there were 200 layoffs.[23] Revenue for the 2002 fiscal year was US$41.1 million, compared to US$26.1 million for 2001. Yet net loss for 2002 was US$130 million, compared to the previous year's loss of US$52.7 million.[24] At the end of the 2004 third quarter, DeCode had US$215.2 million in cash and short-term investments.[25]

Stefánsson is quick to point out that profits from medical research don't just appear overnight. Look at the drug industry, which spends a decade or longer bringing a new product to market. The "new genetics" has not been given the time it needs to deliver, he says, thinking of an analogy: "If you would go to China and get one of those trees that blooms once every 130 years and you put it in your backyard, and someone comes along 60 years later and says, 'You're an absolute failure, your tree hasn't bloomed yet,' how would you feel?"[26] If Icelanders will just give Stefánsson some time, he feels, they won't be disappointed.

Critics of the project question whether Icelanders will ever reap the rewards of their own biological gold mine. They say the people are being exploited by one of their own. Stefánsson's "bioprospecting"—mining the genes of a remote, inbred population[27]—has raised the hackles of ethicists and doctors alike. (For a complete discussion of bioprospecting, see Chapter Nine.) "The resources are no longer metal or spices but medical data and blood for entire nations. The returns will supposedly come later on in the form of free medicine," says one physician. "The rebirth of bartering!"[28]

The sharing of benefits between researchers and subjects is of global concern. According to the UNESCO *International Declaration on Human Genetic Data,*[29] ". . . benefits resulting from the use of human genetic data, human proteomic data [structure and function of proteins], or biological samples collected for medical and scientific research should be shared with the society as a whole and the international community." According to the declaration, benefits may take a number of forms, including access to medical care, new treatments resulting from the research, support for health services, and special assistance for research subjects.[30] The Icelanders' payback for donating their genes? Free drugs.

In 1998, a group of Icelandic physicians and scientists banded together to oppose the commodification of the country's genes. They formed Mannvernd (the name translates as "human protection"), The Association of Icelanders for Ethics in Science and Medicine, and tried unsuccessfully to repeal the Health Sector Database Act. Members encouraged fellow citizens to opt out of the database, but with little success. Patient groups claim that physicians who refuse to hand over medical records have no right to make decisions on behalf of their patients; they argue that the doctors are claiming ownership and control over community property, because the state has financed the recording and assembly of the medical data.

In line with this thought, anthropology professors Gísli Pálsson and Paul Rabinow argue that "[o]nly a paternalistic and rigid hierarchy of knowledge that categorically separates expertise and ignorance, it seems, can accommodate for the contradiction between the Medical Association's concern with relations of trust and their refusal to follow the will of the majority of the Icelandic public."[31]

As for Stefánsson, he says the grassroots organization is totally inconsequential and represents a small percentage of Icelandic society. "Mannvernd claims that I bribed the Prime Minister of Iceland through some shady real-estate deals, that I bribed the political parties of Iceland, that I fooled the Icelandic nation, you know, things of that sort," he fumes. "How dare they?"[32]

No matter how many rivals confront Stefánsson, he always remains standing at the end of the battle. In less than a decade, DeCode has identified susceptibility genes for conditions including obesity, schizophrenia, high blood pressure, asthma, and Type 2 (late-onset) diabetes. In the fall of 2003, the company announced its discovery of a gene that predisposes people to stroke, which takes the lives of 16,000 Canadians and 600,000 Americans every year.[33] Those harbouring the gene have a three to five times greater risk of suffering a stroke than the average person. Hoffman–La Roche is already in the laboratory testing the gene's enzyme, to create future drugs that may be useful in treating all stroke patients, even if they do not carry this specific disease gene.

In November 2003, DeCode announced that it had found its sixteenth disease gene: BMP2 (bone morphogenetic protein 2), which triples the risk of osteoporosis. Scientists combed through the DNA of 1,323 Icelanders from 207 families, comparing those with degenerating bones to healthy citizens. The disease results in around one million fractures annually in the United States, which cost the country US$17 billion in medical care in 2001. DeCode is developing a DNA-based diagnostic test to identify those with BMP2 to minimize bone loss before it begins.[34] DeCode's genealogy database contains records of approximately 95 percent of Icelanders who have lived since 1700, and it has collected genotypic and phenotypic data from a hundred thousand volunteers.

---

Population genetics is not confined to the remote island of Iceland. There is another small country that scientists and physicians have set their sights on. It resembles New England, with forests that cover almost half its land, limestone deposits, and a coastline moulded by countless bays, straits, and inlets. Over the centuries, Germans, Swedes, and Russians have invaded this vulnerable country in the Baltic Sea. When the Russian Empire collapsed in the First World War, the nation gained its autonomy. But in 1939 its independence was quashed when the Soviet Union seized the country, along with Latvia, part of Finland, and Lithuania. Would Estonia be forever a pawn of the great powers?

Then, after centuries of oppression, the Communist curtain was lifted, and on August 20, 1991, Estonia regained its independence. Since then the country has restructured its ailing economy; it now boasts a balanced budget and free trade. The new country is still emerging, with a young, energetic prime minister and an entrepreneurial spirit that is gaining favour in the West. But it is not just investors who are setting their sights on this reformed nation. Researchers are staking their claim to Estonia's people, who live in a considerable degree of geographical and genetic isolation.

In 1999 a group of enterprising scientists, physicians, and politicians known as the Estonian Genome Foundation (EGF) proposed a nationwide genomics project to improve the country's health. The average life expectancy for men is only 72, and the suicide rate remains high despite the improved standard of living. Through genetic studies, the EGF hopes to uncover some of life's genetic mysteries.[35] Genealogical, medical, and genetic data from about a million of the country's 1.4 million people are to be collected. There are many similarities to the Iceland project: data are to be protected by a complex encryption system; medical and genealogical records are linked for gene discovery and drug development; and one company funds the costly venture and has an exclusive commercial licence to profit from the research.

But the EGF is adamant about separating itself from DeCode's controversial consent process. All Estonians are offered the chance to participate, but must agree in writing to do so—the opposite of Iceland's opt-out method. Before they can offer their consent, they must understand what is being asked of them, how their personal medical information will be used immediately and in the future, and what their ownership rights are over their own genetic information. Those interested in participating must visit their physician, who reviews the procedure and the gene donor consent form. Patients are given a bundle of information to read before they are able to sign the form.

For some ethicists, opt-in policies are based on unreal expectations of informed consent, calling into question whether the patient is in fact truly informed. Arthur Caplan, director of the Center for Bioethics at the University of Pennsylvania, says, "In Estonia, there is so much enthusiasm on the part of government officials about this as a way to obtain scientific prominence and some degree of fiscal remuneration, it might be hard to say no."[36] Not to mention the fact that physicians are paid for the time they spend with donors, filling out specialized questionnaires and reviewing the procedure. Approximately 20 percent of Estonia's doctors will be trained to administer the questionnaires.[37]

But maybe North American ideals of autonomy are being unjustly imposed on countries like Estonia. Maybe Canadians and Americans cannot fathom a society that favours itself over the individual, catering to the general needs of the public rather than the unique requirements of each citizen. If personal information is protected, minimizing the possibility of discrimination, why is it so hard to believe that Estonians and Icelanders might want to participate simply to better the lives of their fellow citizens?

"Why are we second-guessing and presuming that everybody who does go for that is dumb?" asks Bartha Maria Knoppers, law professor at the University of Montreal and Canada Research Chair in Law and Medicine. "There comes a point in time when autonomy becomes so artificial, because you're presuming that autonomy is equivalent to, What's in it for me? And that might not be the case. Your autonomous expression of yourself might say, If I'm anonymized, if I'm double-coded, if I'm unidirectionally encrypted like in Iceland, this is it. I'll help my country, help my economy, help find diseases, help future generations."[38]

Knoppers was instrumental in drafting Estonia's Human Genes Research Act, which became law in 2000, to ensure voluntary participation, informed consent, and oversight by an ethics committee. Rooted in the UNESCO *Universal Declaration on the Human Genome and Human Rights*[39] and the Council of Europe's *Convention on Human Rights and Biomedicine,*[40] the Estonian act was endorsed by ethicists and lawyers across the globe. According to this legislation, only gene donors and their physicians may review personal health information; donors have the right to see their own gene-bank records free of charge; no one is obliged to participate in the project; and genetic discrimination is prohibited, with insurers and employers forbidden to use genetic information.

The estimated cost of the Estonian gene bank: US$100 to US$150 million over the five-year period. To help cover the cost, the EGF founded EGeen, a public company based in California. While data are owned and managed by the non-profit EGF, EGeen has the right to pay for access to this information. Its goal is to turn this research into millions in drugs tailor-made for each individual genome, a profitable and life-saving venture.

The United Kingdom is launching its own population genetics project: the UK Biobank. Run by the Medical Research Council (MRC), the Wellcome Trust, and the country's department of health, the Biobank, which is based at the University of Manchester, received initial funding of US$72 million.[41] Its owner is UK Biobank Limited, which holds the right to sell or destroy samples, although it has no intention of selling them. Those who give samples also hand over their property rights to this material. While the UK Biobank has been created for research purposes, this does not preclude the possibility of generating a profit from spinoff inventions.[42]

The Biobank began recruiting volunteers in 2005. Participants must be between 45 and 69, as the project is intended to study diseases that typically strike at middle age, including heart disease and stroke. Approximately 500,000 volunteers will offer lifestyle and diet information, medical histories, and biological samples. There will be an initial assessment, when physicians learn of family histories of disease and present medical conditions. Then, over the next 30 years, this information will be tracked and plotted against participants' ongoing health records, to study the relation between disease, genes, and lifestyle.

An intricate security system protects the privacy of Biobank donors. Health and lifestyle information is stored separately from biological data, linkable only by a code that is extremely difficult to access. Even Biobank researchers handle anonymous samples and data. With this population project, participants are asked for their "informed" consent, even though they have no idea what their personal health information will be used for, by whom, and for how long. A project statement explains, "Since it is impossible to anticipate in advance all of the research uses to which data will be put, UK Biobank will not proscribe any research uses at the outset." Volunteers will not be informed as to which commercial proposals are accepted by the Biobank, and will remain ignorant as to which companies will profit from their genes. They will have to put their trust in the hands of the Multi-Centre Research Ethics Committee (MREC) of the National Health Service, and the Biobank's scientific and ethical review. If a proposal meets set standards, the Biobank has the exclusive right to approve the partnership. When participants give their consent, it is impossible, then, for it to be informed. According to bioethics professor David E. Winickoff and Dr. Richard N. Winickoff, "If individual subjects are being treated with respect, then they understand the purposes for which their tissue or blood will be used, comprehend the risks and benefits of particular projects, and retain the right to withdraw from the study at any time."[43] Clearly, this is not the situation with Biobank.

---

Canada is creating a similar database, which will be governed by the non-profit Institute for Populations and Genetics (IPEG). In Quebec, the Réseau de Médecine Génétique Appliquée (RMGA), which translates as the Applied Medical Genetics Network, has created the CARTaGENE project to map genetic variations in 60,000 volunteers between the ages of 25 and 74. Participants will be chosen randomly to represent the entire Quebec population. CARTaGENE's proposed large-scale genetic and public health studies will allow researchers to home in on genetic mutations that frequently appear in specific regions across Quebec.

An ethics committee will oversee the project, and a toll-free phone line will be used to field questions from the public. CARTaGENE will keep citizens informed through its newsletter, which will be posted on its website. This four-year project is an attempt to create an open relationship with the public, treating citizens more like partners than subjects.[44] CARTaGENE aims to carry out a pilot program in the fall of 2005, expanding to full volunteer recruitment in the spring of 2006.

Such population studies are critical in advancing our knowledge of genetic disease. In different populations, there can be many mutations of the same genes that all result in the identical disease. Thus gene mapping can be a complex, arduous procedure; researchers are not looking for just one specific genetic marker, but must scan countless people's DNA for numerous possibilities. Quebec is home to several homogeneous population groups that exhibit only a few mutations for a particular disease. The BRCA1 gene, for example, has more than 500 variations that predispose women to breast cancer. Yet in Quebec only four have been identified.[45]

The French-Canadian people are ideal subjects for genetic research. It is thought that there were only 8,500 first settlers, including a mere 1,600 women. In the seventeenth and eighteenth centuries they made the voyage from France to Quebec, dropping their anchors up and down the banks of the St. Lawrence River, staking their claim on "Nouvelle-France." The population grew quickly from this isolated group, and the genetic implications are astounding. Since the 1960s, researchers have seen a high incidence of certain hereditary diseases among the Québécois. It has been discovered that 28 genetic disorders are prevalent in at least one region of Quebec, 24 of these in areas east of Quebec City.[46]

Population studies remain a relatively new phenomenon in genetics. But in the not so distant future, these projects will result in exciting findings that could help people in every pocket of the globe. To ensure that such information can be shared across borders, CARTaGENE, the UK Biobank, Estonia's project, and GenomEUtwin, a population database involving eight other countries, have formed Public Population Project in Genomics (P3G). Its mandate: to provide the resources, tools, and knowledge to perfect methods of data management that can be shared by researchers around the world. This international effort is necessary to ensure that all of humanity benefits from the new genetics, rather than selected elite groups, says Bartha Knoppers, the driving force of P3G. When it comes down to our genes, it truly is a small world, after all.

---

For centuries, people have defined themselves by race, religion, and geography. They proclaim their differences by pointing to the colour of their skin and the land they call home. Searching for a sense of identity, nations cling to cultural values and traditions to define their differences from the "other." People spend their entire lives staking out their place in the world, acting as human shields to defend their borders, sometimes even choosing death over amalgamation.

On June 26, 2000, however, scientists offered an alternative human paradigm. The human genome had been mapped. The six billion members of our human species had been traced back 7,000 generations, to a small founding population of 60,000 settlers.[47] Being cut from the same cloth has any two individuals sharing 99.99 percent of their DNA, regardless of how they look, where they live, or what they believe in.

That day, former U.S. president Bill Clinton and U.K. prime minister Tony Blair met at 10 Downing Street to announce the completed first draft of the human genome. Clinton proclaimed, "Today, we are learning the language in which God created life. With this profound new knowledge, humankind is on the verge of gaining immense, new power to heal."[48] Standing by their sides were the two masterminds who had decoded this "book of life": Francis Collins, director of the National Human Genome Research Institute, and Craig Venter, president of Celera Genomics. Together, these two men had painstakingly rearranged the three billion letters—As, Cs, Ts, and Gs—into a readable scroll of DNA, presenting humanity with its own genetic printout. "It has to be a milestone in human history when you have a first look at your instruction book. Having this book will change the world,"[49] says James Watson, who, along with Francis Crick, uncovered the structure of DNA over half a century ago. "Had anyone suggested in 1953 the entire human genome would be sequenced within 50 years," says Watson, "Crick and I would have laughed and bought them another drink."[50]

The Human Genome Project (HGP) almost became a footnote in scientific history. When the $3-billion project was first proposed, scientists scoffed at such an undertaking. Not even the simple bacterium had been mapped. Sequencing the entire human genome seemed a colossal waste of time, energy, and money, not to mention a potential embarrassment of devastating proportions to the entire scientific community if the project flopped. When it was merely an idea being batted about by the U.S. Department of Energy (DOE), geneticist David Bostein denounced the effort as "DOE's program for unemployed bomb-makers."[51] James Wyngaarden, then head of the NIH, echoed this sentiment, saying that proposing such a project was like "the National Bureau of Standards proposing to build the B-2 bomber."[52] Despite vehement opposition, the NIH created the National Human Genome Research Institute (NHGRI) to run the HGP, with Watson as its director.

Yet Craig Venter had his own ideas. The NIH scientist got word of a state-of-the-art computer that uses lasers to identify the chemical letters that spell out DNA. Venter's NIH superiors refused to pay for the prototype, but he went ahead and purchased it anyway. His plan was to use this machine to focus on chunks of DNA that were seen to be actually doing something, rather than attempting to sequence the entire genome, "junk DNA" and all. These targeted bits of DNA would be marked and decoded.[53] Venter said it was just as simple as it sounded, proclaiming that he could find 80 percent to 90 percent of the genes within a few years at a fraction of the cost of the HGP.[54] Watson rejected Venter's approach, but the NIH listened to what Venter had to say. Soon after, Watson stepped down.

Rivalry is not foreign to Venter, who has continually chosen to shun the establishment, rebelling against the goals prescribed by society. After barely achieving a high school diploma in the 1960s, the blue-eyed surfer headed for the beaches of southern California. He was snatched from this sun-and-surf paradise in 1965, when he was drafted for the Vietnam War. Refusing to follow orders, he did two stints in the brig. He served as a medical corpsman and was sent to Da Nang, in southern Vietnam, where he ran an intensive-care ward.

Witnessing so many senseless deaths transformed Venter. He learned how fragile existence could be, and decided to do something meaningful with his life. When he returned to the States, he exchanged his fatigues for a lab coat, earning a Ph.D. in physiology and pharmacology from the University of California, San Diego. In a few years, he found himself at the NIH.[55] Years later, tired of the NIH's penny-pinching and cost-cutting, Venter left with another NIH scientist—his wife, Claire Fraser—to launch The Institute for Genomic Research (TIGR). The private firm Human Genome Sciences (HGS) would partner with TIGR to profit exclusively from Venter's research, with Venter also cashing in on the partnership. Now science had a personal edge; Venter would race his old colleagues to map the entire genome.

After Watson stepped down as director of the NHGRI, Dr. Francis Collins took the helm, riding in on his motorcycle, wearing a helmet plastered with stickers of the genes he had already mapped. In addition to mapping the cystic fibrosis mutation with Lap-Chee Tsui (see Chapter Two), Collins had helped Nancy Wexler and her team peg the Huntington's gene. But Collins and Venter were diametrically opposed in their upbringing and perspectives on life. Collins had grown up on a 95-acre farm in Virginia's Shenandoah Valley. His parents, Margaret and Fletcher Collins, had taught their son the importance of a good education. His father had a Ph.D. in medieval literature and his mother wrote plays that were performed in a tiny theatre nestled in an oak grove on the family farm. At age seven, Collins wrote a script based on *The Wizard of Oz* and directed the play himself. He was home-schooled by his mother until he was 10, graduating from high school just six years later.

Studying biochemistry at Yale University while working on his Ph.D., Collins was introduced to DNA's life script. He says, "I was completely blown away." After graduating, he secured a junior faculty position at the University of Michigan, where he stayed until 1993. The chance to spearhead the HGP, science's most adventurous and daunting endeavour, lured him away from teaching. Saying goodbye to his class, the scientist reached for a well-loved guitar, his cowboy boots peeking out from under his lab coat. He belted out his own rendition of a Frank Sinatra favourite. "So start today," he sang, "love DNA, and do it ouuuurrr way."[56]

It is hard to believe that, despite their intense rivalry, Collins and Venter were allies who jointly accepted credit for mapping the human genome. Author Ingrid Wickelgren compares the scientists to a famous pair of duelling Second World War generals. Collins is Britain's General

Bernard Montgomery, who carried out methodically planned attacks, while Venter is U.S. General George S. Patton, who blindly forged ahead even when told to stand his ground.[57] As Patton liked to say, "A good solution applied with vigor now is better than a perfect solution applied ten minutes later."[58]

While Collins and the HGP team had been systematically mapping different sections of DNA along chromosomes before sequencing them, Venter and the TIGR scientists threw down the gauntlet with the "whole-genome shotgun" (WGS) approach. Venter randomly tore off millions of pieces of DNA that he immediately sequenced and fed into his supercomputer, which then fit the pieces back together by referring to overlaps in the sequence. Collins announced that such a map would be so abridged and error-ridden that it "would read like Cliffs Notes or *Mad* magazine."[59] Yet in mere months, Venter had published the bacterium genome, sequencing for the first time the DNA of a living organism.

While Venter enjoyed his new-found wealth from his partnership with Human Genome Sciences—he sold his equity holdings in the company for more than US$9 million and bought an 82-foot yacht—money couldn't mend his relationship with HGS CEO William Haseltine. Haseltine considered Venter a mediocre scientist, while Venter likened Haseltine to a slick Gordon Gecko–style financier.[60] In 1997, Venter walked away from his contract. "I gave up $38 million to get away from [Haseltine]. It was the best $38 million I ever spent."[61] But Venter needed money—and lots of it—if he was ever to pull the trigger on his shotgun approach.

Enter health sciences company Perkin-Elmer Corp. and its president and CEO Tony White, who, along with Venter, launched Celera Genomics to fast-sequence the genome. Embedded in the word acCELERAtion are the letters that formed the name of Venter's new company, suggesting speed. White offered US$300 million to jump-start the HGP competition. Venter had the PRISM 3700 machines to shotgun the genome, and even offered to let government researchers use them after the project was completed. Speculation heightened as to the purpose of using taxpayers' money to fund a separate sequencing effort, when Venter seemed to have the whole thing wrapped up. Venter boasted that he could sequence the entire genome by 2001 rather than the proposed 2005, at a tenth of the cost—US$300 million instead of US$3 billion. Rather than give up after a decade of work, Collins and the HGP team ramped up their efforts, and the race continued.[62]

Using 300 Perkin-Elmer sequencing machines that cost US$300,000 apiece, Celera tested its sequencing method by tackling the genome of the fruit fly *Drosophilia melanogaster.* The machines spit out the fruit fly's genetic script, and Venter proved that his method was worth millions. Then came the war of the press releases. In October 1999, Venter announced that Celera had sequenced one billion bases of the human genome. The NIH countered the announcement with its own: to celebrate the HGP sequencing of one billion bases, the NIH was hosting a "birthday" party at the National Academy of Sciences, with scientists donning T-shirts splashed with the double helix. In January 2000, Venter declared that they had sequenced 90 percent of the genome. In March, the HGP boasted that it had sequenced two billion bases.[63] Venter and Collins continued to one-up each other, until word came from the big white house on Capitol Hill.

In April 2000, President Clinton was tired of the scientists goading each other on like little boys in the schoolyard, and told scientific advisor Neal Lane, "Fix it . . . make these guys work together."[64] Ready to win the favour of the scientists who had denounced him for profiting from the genome, Venter was willing to talk. It probably didn't help that millions had been lost almost overnight. When British Prime Minister Tony Blair had announced that genomic data should be free, Celera's stock had plummeted from US$189 a share to US$149.25. Neither scientist wanted their personal feud to overshadow what was being hailed as science's greatest achievement, which some said would eclipse the Apollo mission that had landed Neil Armstrong on the moon.

Venter and Collins met comradely over pizza and beer, and by May of 2000 they had agreed to accept joint credit for mapping the human genome.[65] The pair had also settled when they would publish their data. Celera's paper appeared in the February 16, 2001, issue of *Science,* and the HGP results were published a day earlier in *Nature*. With the publication of Venter and Collins' findings, humankind's lofty vision of itself was somewhat deflated. The human genome has approximately 35,000 genes, rather than the once-estimated 80,000 to 150,000. It turns out that we have only double the number of genes than that of a worm or fly.

On January 22, 2002, Venter resigned from Celera. The scientist who had left the NIH a decade earlier with only US$2,000 in his bank account boarded his multi-million-dollar yacht to sail around the Caribbean.[66]

---

In November 2003, humanity was given the final draft of its genetic printout. While the rough version had been riddled with errors, the chance of error had now been reduced from one mistake in every 1,000 base pairs to one in every 10,000. Scientists continue to pore over Venter and Collins' astounding effort, trying to grasp what makes us human. Gaining this staggering sense of self-awareness, though, is like reverting to when life first began. We are not all that we envisaged ourselves to be, and we can be genetically decoded with the effortless click it takes to run a computer program.

Our genes have been plotted and mapped, but the big mystery remains: what do our genes do? How do they churn out the proteins that grow our hair and nails, produce insulin, and combat infection? What switches certain genes off while turning others on? What are the genetic prompts that activate disease? Scientists say the answers to life's biological riddles can be found in the smallest genetic variations between people. While most of our genetic makeup is identical, it is the 0.1 percent—our individual genetic differences—that confer risk of disease or affect our response to medication. The most minute genetic variation can reveal our biological destiny.

When genetic sequences vary by a single base, they are called single-nucleotide polymorphisms (SNP). For example, some people may have a G base on a chromosome in the place where others have a C base. A person's entire index of genetic variations is known as his or her genotype. Since Gs and Cs like to cling to each other, someone could have CC, CG, or GG. If an SNP is inherited along with a disease, it is believed that the disease-susceptibility gene is hiding right around the corner. To discover a person's SNPs across the entire genome is called genotyping.

Every genome has common genetic variations, called "tag SNPs." Geneticists believe there are approximately 10 million common SNPs nestled throughout the human genome, but it is technologically impossible to scan for all these variations. So scientists are focusing on sex cells to guide their search. When cells divide, chromosomal pairs meet and swap genetic codes, creating blocks of genetic material that are inherited as a whole from one generation to the next. These blocks are called haplotypes. There are common haplotypes found throughout the genome, which can be uncovered by studying a few important SNPs. It is thought that there are only a few typical haplotypes in different populations.

The International HapMap Project aims to create a map of these haplotypes that could uncover the complexities of disease and individual responses to medication. Scientists from around the world are collaborating to map the remaining 0.1 percent of the genome, which many predict will be far more revealing than the other 99.999 percent. This three-year, US$100-million effort will study 270 samples from Northern European and Western European populations, a Nigerian people called the Yoruba, and Japanese and Han Chinese populations.

The initial goal is to study 600,000 SNPs hiding throughout the genome. In line with the Human Genome Project, the creators of the HapMap intend to release data quickly and freely to the public. There will be no fee to view research results, which will further the ongoing study of genomics.[67] "The HapMap will provide the missing link between the DNA sequence of the genome and the way in which the genome influences the risk of disease," says Collins, who is now director of the National Human Genome Research Institute, which is helping to fund the project.[68]

HapMap genotyping will be done at research facilities in Canada, the United Kingdom, the United States, Nigeria, Japan, and China. Each team will be responsible for scanning a designated chromosome. Canada, for example, is in charge of chromosome 2 and part of chromosome 4, which amounts to 10 percent of the entire project. The Canadian arm of the project is funded by Genome Canada and Genome Quebec, for a total of $15 million. Genotyping is to take place at McGill University and the Genome Quebec Innovation Centre, under the guidance of geneticist and immunologist Dr. Thomas Hudson. Personal information of the participants will not be linked to samples, making it impossible for individuals to be identified; samples will be labelled only by population and sex markers. The research will focus solely on genetic makeup, and will not include medical information, as the population databases do. Yet there is concern that HapMap results may stigmatize the targeted populations if they are found to be susceptible to certain conditions.

Over three decades ago, groups of people were already being genetically marked and discriminated against. In the 1970s, scientists perfected a test to identify carriers of the gene involved in sickle-cell anemia. The U.S. Air Force Academy concluded that since the gene was responsible for hemoglobin, which carries oxygen within the blood, such carriers could not perform work functions at high altitude, where lack of oxygen was already an issue. The majority of people with sickle-cell anemia are black. The Academy banned carriers until protests of racism forced it to lift the restriction.[69]

To minimize such discrimination, the HapMap consortium will educate the public and researchers on genetics, and the realities and uncertainties of genetic results. Researchers will be taught how to share their findings to create a well-informed public. Samples do not represent the entire populations they are being taken from; for example, the Yoruba samples are not archetypical of all Africans, or even West Africans. In addition, people should be taught not to equate "genetic" with "unchangeable," states the NIH, and should not have the impression that "because someone has a particular genetic variant they are 'doomed' to get the disease."[70] People should know that lifestyle also affects disease onset. The goal of the genomics era is to normalize and integrate genetic information, says Bartha Knoppers, adding that we already went through this with cancer and psychiatric disorders, and now it is time to do so for genetic diseases. If we are all at risk, Knoppers envisions a greater equality arising from this knowledge.[71]

With the human genome now mapped and the HapMap in progress, pharmaceutical researchers are working in the lab, comparing the genetic makeup of diverse people, to create genetic tests that will predict individual responses to medications. This science is called pharmacogenetics. Individuals vary greatly in how they process medication. Our bodies metabolize and get rid of drugs differently. And some individuals are sensitive to drugs, even in low doses.[72]

Drug companies envision tailor-made treatments for patients who are not responding to their current prescriptions. In the U.K., 1,100 people died from adverse reactions and prescription errors in 2002, which cost the country's health care system more than £500 million. In the United States, there were approximately 400,000 adverse events from prescription or over-the-counter drugs in 2001.[73] Underdosing, overdosing, and incorrect prescribing cost America more than US$100 billion a year in hospitalization, premature deaths, and reduced productivity.[74]

The HapMap may show us that some SNPs live in the area of a gene that is involved in making proteins. The amount of protein a gene produces can directly affect the way the body responds to certain drugs. If an SNP is the neighbour of a genetic variation that alters a person's metabolizing of a drug, the neighbours are typically inherited as a pair—so SNPs can be used to pinpoint the genetic variants targeted by pharmacogenetics. The genetic variation of the aldehyde dehydrogenase (ALDH) gene, for example, affects people's reaction to and absorption of alcohol. Those with a particular variation of ALDH have difficulty processing alcohol. They feel nauseated, their faces turn red, and they get a splitting headache.[75]

If the body produces a high volume of the HER2 protein, it affects a woman's chance of developing breast cancer. A study published in *Science* magazine found that "25 to 30 percent of patients with breast cancer express high amounts of HER2 due to a somatic mutation in the DNA of the cancerous cells, which contain many copies of the HER2 gene."[76] Patients have an increased chance that the cancer will spread and that the disease will resist chemotherapy, and they could expect a shorter lifespan. The big pharmaceutical company Genetech created Herceptin, a drug that treats this specific group of women by targeting overexpressed

HER2. To determine candidates for the drug, physicians measure the HER2 in a woman's tissue.[77]

Now that our genetic scroll has been rolled out for all to read, life will forever be changed. We can now peer into ourselves and into our own unique biological destiny. As James Watson muses, "We used to think our future was in the stars. Now we know it is in our genes."

## Chapter Four

# Body Language

### Interpreting the Messages of Stem Cells

The hospital never sleeps. In the dead of night, when storefronts have gone black and porch lights have been flicked off, frazzled parents pile their sick children into back seats and navigate through dark neighbourhoods, the fluorescent H on their radar. Cars pull in front of the blinding emergency sign, and are abandoned. The sliding glass doors whoosh open, sucking families into an obstacle course of endless corridors, gurneys, IV poles, and groggy people waiting. They come here looking for answers, and everything else—time, hunger, sleep—is forgotten.

The Ottawa General Hospital exists in this timeless orbit. Visitors needn't leave for days. There is the food court, the bank machine and gift shop, even information desks for those lost in the maze of dulled white hallways. But one corridor leads to a different part of the hospital. Follow the trail of white coats to the bank of elevators and up to the fourth floor. Beyond a locked door is another blank hallway, but there are no patients here. Instead of nurses watching over hospital beds, there are scientists tending microscopes. A sunlit laboratory is bursting with shelves of beakers and test tubes, and tables cluttered with the day's experiments.

In a cramped office in one corner of the lab, Michael Rudnicki leans back in his chair. A comfortable shirt is tucked into his relaxed black jeans. He stretches out his long legs and turns away from his computer screen. Between his thumb and index finger he twirls a computer chip that holds the answers that those exhausted parents down in the ER have come in search of. Affymetrix's GeneChip contains half of the human genome. When Rudnicki stacks two chips in his palm, he is grasping the clues to life's genetic riddles. With the genome now mapped, the scientist can identify our approximately 35,000 genes. But now comes the hard part—figuring out what they all do.

"From a scientific point of view, understanding what makes these cells tick is endlessly fascinating," admits Rudnicki, the director of Ottawa's Centre for Stem Cell and Gene Therapy.[1] If he can map the mechanical blueprint of stem cells, sketching how their chemical pathways are activated and how they compel cells to react, he can identify targets for therapy. Once he knows what proteins are secreted in patients with certain diseases, medication can be tailored to reverse the negative effects of these proteins, encouraging the stem cells to function normally.

Since he was a boy, Rudnicki has been captivated by the intricacies of life. His voyage into science began in his Ottawa home, where he watched National Geographic specials and dreamed of becoming the next Jacques Cousteau. During family trips to a local lake, he used to splash around in snorkel gear. One summer he combed the Nova Scotia seashore for creatures, finding it endlessly fascinating. Rudnicki's mother worked as a nurse and later transferred to social work. His dad was also in social work. Rudnicki has two sisters and two brothers—one of them his twin, an electrical engineer and the only other sibling to go into science.

Rudnicki gave up marine life for human biology while working on his undergrad degree at the University of Ottawa. One of his professors, David Brown, was a "real experimentalist" who pushed students to ask questions to figure out how life worked. Rudnicki was lured into this biological puzzle, and spent years trying to fit all the pieces together. After completing his Ph.D. in 1998, he then trained as a post-doctoral student at MIT's Whitehead Institute, where he continued to study genes. In 1992 he accepted his first professorship, at McMaster University in Hamilton, where he stayed for almost nine years. Dr. Ron Worton, the director of the Stem Cell Network, lured him back to Ottawa, where he continues to study the intricate workings of genes.

While many geneticists remain focused on gene therapy, Rudnicki believes that our healing powers lie within our stem cells—the body's most primitive cells. He feels there is no need to taxi new genes through our biological hills and dales; so much can go wrong along the way. He believes that we harbour the materials needed to repair our own creaking bones, failing organs, and lifeless tissue—that we can be our own specialized pharmacy. And Rudnicki is intent on writing us a prescription for health and well-being.

---

Stem cells are virtually impossible to identify. They don't look any different from their neighbours, and there is only one stem cell in every 100,000 cells. But these rare cells possess the power to create life. Stem cells are born as blank slates, free of any specific destiny or purpose. One day they may become muscle, blood, bone, or almost any other component of the human body. But for the first few days of their existence, their only function is to make carbon copies of themselves. Scientists are fascinated by their ability to divide again and again while remaining totipotent—"totally potent"—capable of anything but committed to nothing. The body houses two types of stem cells: embryonic and adult. When an egg is fertilized by a sperm and an embryo begins to grow, cells start to divide, creating totipotent cells. By day four these cells cling to each other to create a blastocyst, which will metamorphose into an embryo, and if left to its own devices, a baby. The outer layer of the blastocyst is made up of cells that will develop into the placenta. They are called pluripotent—"plurally potent"—meaning that they have the ability to transform into almost all of the body's cell types, but are not as powerful as their totipotent sisters and cannot create a complete living organism. The blastocyst's inner cell mass also has pluripotent cells that can morph into most of the body's tissues but, again, not a person.

Also inside the four-to-five-day blastocyst are embryonic stem cells, which can become any of the three layers of the embryo: the inner, middle, or outer. They express a protein called Oct-4, which targets genes that instruct the cells to remain proliferating and undifferentiated for long periods of time. If these cells could be grown indefinitely, and could eventually be programmed to turn into a desired cell type—say neurons—they could be used to replace ailing cells in Parkinson's or Alzheimer's brains. If a patient's heart was failing, stem cells could be transformed into muscle cells to replace damaged heart cells and get the heart pumping again. And stem cells could renew themselves and become a type of bone marrow cell, to rejuvenate the blood of leukemia patients.

While embryonic stem cells are found within the blastocyst, adult stem cells are found within a specific type of tissue in the body. They call many organs home—the brain, pancreas, liver, and even the skin and the retina of the eye. While adult stem cells begin as undifferentiated cells within a certain tissue, once they duplicate and renew themselves they quickly become one of the specialized types of cells of their home tissue. This process is called self-renewal. These adult cells can pump out future generations of specialized cells, but their scarcity makes them hard to rely on for treatment. And as far as scientists can tell, a line of adult stem cells will not create new generations indefinitely in culture; eventually they come to "the end of the line" and are of no further use.

---

But what makes a stem cell a stem cell, asks Rudnicki, who is studying these coy cells to determine their path in life. How is it decided what they will be when they grow up? The Stem Cell Genomics Project is a $10-million project that studies "expression profiling" in adult and embryonic stem cells and their daughter cells that have differentiated (developed into muscle, bone, or other cells). The Stem Cell Network—one of Canada's Networks of Centres of Excellence, which support Canadian research—contributed $1 million in seed funding to Rudnicki's project, with Genome Canada giving $5.6 million and additional funding coming from other organizations. After seven years, Rudnicki, who began the project in the spring of 2003, will have to reapply to the Stem Cell Network for further funding.

A GeneChip is a computer chip that allows Rudnicki to scan 35,000 samples simultaneously, which will let him identify patterns in gene expression. Before the era of the GeneChip, he did expression profiling one gene at a time. His samples include hematopoietic (related to blood cells), embryonic, neural, and pancreatic stem cells from mice and from humans. During the course of the study, Rudnicki will analyze 1,500 chips. Within each chip, which costs $1,000 to evaluate, there are numerous stem cell genes for him to cultivate.

The project is different from other stem cell ventures. "Primarily what investigators have done is isolated cells and, using descriptive tools, described the function of cells. But they haven't understood how [the cells are] going about that function," says Rudnicki, adding that this is of vital importance in understanding cellular mechanisms. Genes have a hierarchy of control. He compares them to power switches in a house. Imagine the electrical wiring that leads from a house's power feed to every light in every room. Similarly, a series of genetic switches leads down a particular pathway from one gene to the next. With each switch a cell's choices become more limited, as it takes its pathway to a differentiated state. If Rudnicki uncovers which genes activate faulty proteins, future therapy could be as simple and painless as a daily pill or injection. As William Haseltine, former chief executive of Human Genome Sciences, says, "When we know, in effect, what our cells know, health care will be revolutionized, giving birth to regenerative medicine—ultimately including the prolongation of life by regenerating our aging bodies with younger cells."[2]

The body has an innate ability to regenerate itself, says Rudnicki. "But it's something that becomes dysfunctional as we age, unfortunately. So small children who are in horrific car accidents can heal and be perfectly normal adults by the time they grow up." But, he says, "If an adult has the same car accident, they remain a cripple." Humans have a "newt-like" ability to regenerate. If a surgeon has to remove three-quarters of a patient's liver, the organ will grow back. Of course we cannot regrow our limbs, "but if you lop off the tip of your finger, below the joint, maybe halfway through the nail—in a small kid that will almost completely regrow. In an adult that will regenerate, but to a lesser extent." Yet, he says, a fundamental question remains: "What happens during aging to stop that regenerative ability, and are those mechanisms still deployable?"

Down the hall from Rudnicki's office, a team of scientists are exploring life's genetic determinants. Their goal is to identify the genes that either maintain stem cell identity or persuade cells to differentiate into a committed type of cell. This laboratory houses the GeneChip Workstation, a $350,000 machine built by Affymetrix that scans massive amounts of genetic material at a time. Its job is to measure the expression of a gene in countless human and mouse samples, to establish which genes are highly expressed in stem cells. The GeneChip has a window the size of a fingernail that is lined up in the Workstation and scanned for genes. Genetic scanning is a two-step process. First, the entire genetic pattern is captured, similar to the taking of a fingerprint. Then the fingerprint is analyzed to determine which section of the print compels a stem cell to act. The machine's computer program converts the image into numbers. The more highly a gene is expressed, the stronger the signal will be from this data. Samples are compared, plotted, and analyzed by a bioinformatics team. (Bioinformatics manages and analyzes biological data.)

The laboratory is wired to a roomful of computers that download and make sense of the data. A group of programmers and bioinformatics Ph.D.s have created Stem Base, a database of genetic samples. Samples have been annotated so scientists can track the genetic material for future projects. Scientists can log onto Stem Base and ask specific research questions. They can review the results from experiments done on the cell type they are interested in, and compare what they have found to past conclusions of their colleagues. Scientists can mine the data to trace interesting patterns in gene expression in certain stem cells, and follow the process of differentiation from stem cells right through to cells with specific functions. The life of an embryonic stem cell might be followed along a series of changes over time, from zero hour to, say, 12 hours. Data can be used in many different ways, to explore fundamental questions of stem cell identity.

Bringing research back to the patient is the primary goal. Researchers can now identify cells that suddenly go haywire and turn into cancer cells. But how can this knowledge be used for therapy? "We always want to go back to the bedside," says Pearl Campbell, director of the laboratory. "We can understand things in a lab, but it's really of no utility until you take it and associate it to a person who has a disease."[3]

In the mid-1980s, scientists drastically changed the lives of patients with chronic renal (kidney) failure by perfecting recombinant human erythropoietin, or EPO, a naturally occurring hormone that binds to a receptor on the surface of the target cells and tells them to make red blood cells. Anemics no longer had to endure repeat transfusions; an injection of EPO would boost their ailing blood systems. Unfortunately, athletes determined to turn themselves into superhuman workhorses have abused the drug. An elevated production of red blood cells can increase the oxygen flowing to muscle tissue, heightening physical endurance. EPO hit the cycling scene in Europe like a plague, according to *The New York Times Magazine*. "There were police raids, huge stockpiles of EPO confiscated from cyclists' hotel rooms, arrests, trials, wholesale suspensions of competitors. 'Each racer had his little suitcase with dopes and syringes,' a former doctor for European professional cycling teams told a British newspaper. 'They did their own injections.'"[4]

Tricking a healthy body into producing excess red blood cells can result in death, but EPO is an ideal remedy for many patients, including heart attack and stroke victims. With a quick injection or the gulp of a pill, stem cells can be taught to make red blood cells or many other types of cells the body may be lacking. Cellular treatment like this is the future of regenerative medicine, says Rudnicki; a variety of stem cells could one day be targeted by drugs and mobilized. Developing drugs that target stem cell populations would revolutionize the treatment of disease.

Rudnicki is studying another substance found within the body that could renew degenerated muscle. PAX7 turns genes on and off. If the body does not have enough PAX7, which Rudnicki calls the genetic switch, certain genes involved in building muscle are absent. But there are stem cells hiding in pockets of muscle, waiting to be activated. When they are given the go-ahead, they develop into "muscle progenitor cells" that make up 3 percent to 5 percent of all muscle cells and, says Rudnicki, are believed to be the primary mediators of regeneration. These so-called satellite cells sit beside muscle fibres, and if the progenitor cells are inactive, muscle is unable to grow. Once the progenitor cells have divided around eight times, they bind together to form a muscle fibre. Rudnicki has discovered that Wnt is the signal that expresses PAX7. If he can only come up with a drug that replicates this signal in patients with dormant Wnt, it will be possible to bring the muscle's sleeping stem cells to life. Such therapy could help those with muscle-wasting diseases like AIDS, multiple sclerosis, and muscular dystrophy.

Even though future clinical applications are still in the theoretical stages, "Certainly we hope that by learning how to activate the Wnt pathway in an appropriate way we will be able to improve tissue regeneration," says Rudnicki. And the only way to expand the therapeutic uses of adult stem cells, he says, is to continue to study and uncover the mechanisms that decide cell fate.

---

What if, contrary to past scientific belief, the fate of adult stem cells is not sealed? What if these cells are malleable like their embryonic siblings? As science has often illustrated, many "certainties" can be disproved. And this, it seems, is one of them. Over the past several years, scientists have shelved textbooks that held that adult stem cells were fixed in function, and have brought these elusive cells back to the lab. It now appears that adult stem cells can recast their future. They have astounding plasticity after all, and can transform into cells of completely diverse types of tissue. From a biological and regenerative perspective, this is profound, says Jane Aubin, professor and chair of Anatomy and Cell Biology, and professor of medical biophysics at the University of Toronto. This revelation opens up amazing possibilities. Stem cells could become insulin-producing islet cells to treat diabetes, or dopamine-secreting neurons to reverse the effects of Parkinson's.

Aubin, who has cropped greying hair and is professionally dressed in a black pant-suit, is heading the Stem Cell Network's Plasticity Project to uncover how adult stem cells alter their identities, beginning as one type of cell and turning into another. What could switch them from, say, bone to blood or nerve? Aubin's excitement over this two-year, $1.07-million project is infectious. She mesmerizes students and colleagues with the regenerative possibility of stem cells; the mere mention of these elusive little cells puts an extra twinkle in her eye. Aubin possesses that special professorial gift of making learning fun. She spends hours discussing the promise as well as the frustrations of stem cells, drawing listeners in with outstretched, emphatic arm gestures. The tedious experimentation of science can be a challenge, she says, adding, "Once you get it to work the way you want, I think there's almost nothing like it. Maybe for people who play the stock market, when their stock goes up, maybe they get the same buzz."[5]

Born in 1950, Aubin did not always envision a life in science. As a kid, she daydreamed about leaving her hometown of Cornwall to travel the world as a French translator. "I thought it was an incredibly exotic thing to do, to translate from one person to another." Her French-Canadian father worked at the post office. When she was 13, he was transferred to Smiths Falls, a small town in Eastern Ontario. She moved along with her parents, five brothers, and three sisters. She learned from her grade-nine French teacher that hers was one of only two French-Canadian families in the English-speaking town. Throughout her childhood, she says, French "was always part of what you were and what you heard. There's a fun side of French culture. I just thought, This is terrible."

In high school, Aubin's math and chemistry teachers ignited her spark for science. She went on to do her B.Sc. in both subjects, and continued in these fields for her Ph.D. at the University of Toronto. While doing graduate work in medical biophysics at the Ontario Cancer Institute, she became swept up in the enthusiasm over stem cells. It was here, years earlier, that Dr. Ernest McCulloch and James Till had identified the first stem cells while studying bone marrow in mice. They had revealed to the world that hematopoietic stem cells had the power to multiply by the millions, as well as to differentiate into all types of blood cells.

Before settling in Toronto, the city of stem cells, Aubin moved to West Germany to do her post-doctoral work in cytoarchitecture ("cyto" means "cell"), analyzing what causes a cell to achieve its shape and gain an ability to move. Living near the East German border in the late 1970s was an awakening experience for her. East and West Germany were separate worlds, and police marched the streets with machine guns. Aubin now reflects, "I got to see the consequences of having trenches and a big huge fence running down the middle of your country." In West Germany, science had its own oppressive hierarchy. "There's one big head honcho with a team under him—it's normally *him,*" she says. "Even now, there are fewer female head honchos in Germany."

Aubin left the constrained German atmosphere and returned to Toronto, where she continued to explore cellular issues. During her research, she thought back to the groundbreaking work of McCulloch and Till, which inspired her to focus on the mother of all cells. Her team of 17 stem cell specialists from eight Canadian universities are now testing the plasticity of adult stem cells by isolating them from one tissue in the body, culturing them in dishes in the laboratory for a few weeks, and coaxing these committed cells to change their nature and become something completely different. Every scientist focuses on a certain cell type. There are bone, neuron, even blood specialists, all with their own "recipe" for growing specific cells. So if the bone specialist wants to turn her cells into neurons, she uses the neuron specialist's recipe to transform her cobblestoney bone cells into wispy, fibrous nerve cells. With such collaboration, research can progress much faster and more efficiently. "If you start observing your bone cell turning into a muscle cell," says Aubin excitedly, "you have the experts who can raise, and force you to consider, all the possible mechanisms."

If adult stem cells could be multiplied by the millions, and then given a recipe to turn into a completely different cell type, our whole approach to medicine would change radically. Large amounts of cellular material would be convenient and accessible, and rejection would not be a problem because patients would be receiving transfusions of their own cells. The problem we have to overcome is that adult stem cells are unpredictable. They multiply only as much as they please, and begin differentiating without warning. If we cannot consistently grow millions upon millions of cells, we may have to continue relying on embryonic stem cells. But Aubin feels the case is not yet closed. While stem cells have this tendency to differentiate, she says, "It could just be that we haven't got quite clever enough to figure out what would keep adult stem cells also multiplying indefinitely," like their embryonic sisters. Scientists may not have perfected the conditions for these adult cells to divide and multiply. Or could it be that there is a biological clock in adult stem cells that is set once life begins, that scientists can never reprogram?

It has also been debated whether a living organism begins as one single stem cell. Rather than having separate embryonic and adult stem cells creating every tissue in the body, maybe these cells are all related. "The possibility is, a stem cell is a stem cell, and it's there and we just didn't know it," says Aubin. "That the same kind that sits in the embryo, that can do everything, are put out during development into other tissues of the body. It would be a wonderful streamlining of a biological potential, but we don't know that yet." The reality is that scientists know far less about adult stem cells than about embryonic ones, and will have to continue to research both until there are more certainties.

Aubin also poses a "plasticity question"—where do adult stem cells end up once they are released back into the body? Do they take on the personality of the tissue in which they fall? "The plasticity question is," she explains, "does a stem cell go to the blood and make some blood, does it go to the brain and make some brain?" Stem cells might also adopt new functions simply by fusing with their new neighbours. Rather than changing their own characteristics, they might fuse with cells that are functionally integrated into the tissue. "It isn't just, Can we prove cells have plasticity?" says Aubin. It's "Can we prove they have plasticity other than fusion?" A cell changing its own fate is far more fascinating than one fusing with another cell and copying its every move. Then again, once inside the body, the stem cells could also choose to do nothing at all.

Identifying stem cells remains a tiresome and arduous task. Trying to figure out what the cells are doing and why is equally laborious. Embryonic stem cells are easier to locate, because scientists know where to look during development. Adult cells can escape even the most trained scientific eye. "Just seeing something sitting in a tissue doesn't tell you it's functional," says Aubin. "It tells you it's sitting there. And it can live there for ages and ages and not actually contribute, in any regenerative-medicine sense, to the health of that tissue."

Even if Aubin and her team can uncover the nature of adult stem cells, the puzzle remains how to teach these cells where to go. As with gene therapy, the targeting of a specific tissue or organ is tedious and often unpredictable. Stem cells roam around the body haphazardly, as if by accident, before finding a home. Maybe, says Aubin, these cells need to find a "lock and key" with surface proteins on other cells. Even though it has been over 40 years since Dr. Ernest McCulloch and James Till discovered the first blood stem cells, scientists continue to uncover more questions than answers. Aubin says the pair were as much in the dark as to what stem cells were for blood, as scientists are today as to what stem cells are for in the rest of the body.

---

In 1958 the Ontario Cancer Institute opened its doors in downtown Toronto. The tidy red-brick building had beds for 100 patients, outpatient facilities, and laboratories for cancer research. It was a unique institution that encouraged collaboration between chemists, physicists, biologists, and physicians. Clinicians and scientists worked together in the areas of cancer diagnosis and treatment, and the study of the causes of cancer, in an attempt to better understand one of our most perplexing diseases. But people didn't want to be admitted to a cancer hospital, so the facility was renamed the Ontario Cancer Institute/Princess Margaret Hospital after a chance visit by Princess Margaret.

It was estimated that the institute would care for 3,000 patients but it ended up with three times as many, says Dr. Ernest McCulloch, a balding, quiet man of small stature whose thoughtful gaze surveys the world through large round glasses.[6] His thin, straight mouth reflects a reserved nature, giving one the distinct impression that he wants to be left alone to methodically research life's biological quandaries. Born into a family of doctors, McCulloch grew up in Toronto, attending Upper Canada College, an elite private school that he says had only one good science teacher.

"The rest of their science," he says, "was dreadful." It wasn't that McCulloch dreamed of a life in science. Rather, he systematically reviewed his options, and decided his life's path. "My father was a doctor. Two of my uncles were doctors," he says matter-of-factly. "So I went to medical school." He explains that "in the kind of family that I was brought up in—it was a professional sort of family—it was either the church, medicine, or law. I think my father wanted me to go into the church, and I didn't have any inclination that way at all. . . . I looked at law, and what worried me was that it was man-made. So it seemed to me that if you studied law in Canada, it might not work in the United States." But if your study involved people, "no matter where they came from, a person is a person."

After high school, McCulloch sped through his university studies, receiving a medical degree from the University of Toronto at the young age of 22. He was awarded a fellowship to do research in London, England, for being the top student in third year. It was a great honour but it didn't pay much. As luck would have it, McCulloch's father had some money in a British bank. He told McCulloch to "accept the damn fellowship and go spend my money in London," which the young man did, immersing himself in the world of genetics.

McCulloch was awakened by the vibrant city. "I found out how limited our education is [in Canada]," he says. "The people I met in England were broadly educated in a way that I wasn't. They could recognize art. They would always suggest books for me to read. They understood music, understood its value. I brought back a realization that I was pretty ignorant. Still am."

McCulloch returned to Toronto and continued his medical research, branching out into the sphere of pathology—the science of disease—which he says is the root of all medicine. He also put in his time as a hospital intern, albeit against his will. "It became evident that I couldn't fight against the stream," he admits. "I had to do what I was told, like everybody else." Then he was invited to join the Ontario Cancer Institute. James Till, a recent Ph.D. graduate from Yale University, also came on board. It was the 25-year-old's first formal research position. McCulloch remembers handing Till a note to read. According to McCulloch, Till looked at this and said, "Your parents spent an awful lot of money to have you taught to write badly. Jim Till went to a small school out west and writes beautifully."

In many ways, Till is the opposite of McCulloch. He is tall and boisterous and always moving. He is full of life. When he gets wrapped up in his own animated conversation, which he often does, he punctuates his points by shaking his hands and diving at you excitedly. "I come from Western Canada, with an agricultural background, and went to a school that, shall we say, wasn't considered the best in the nation," he says. And McCulloch "was from a medical family in the biggest English-speaking city in Canada and was a graduate of Upper Canada College. Quite a different background."

Till is from the small town of Lloydminster, which straddles the border between Alberta and Saskatchewan. His dad was a wheat farmer, and Till thought he might go into farming until "I found I could make a living without having to do physical work," he laughs, recalling the back-breaking chores of harvest time. One year he helped the threshing gang in the fields. "They were very nice," he says, in giving him the best team of horses. "All you had to do was cluck your tongue and they'd move up one length of the bundle rack."[7] The farmhands effortlessly heaved massive bundles of wheat onto the pile, but Till could lift only one or two sheaves at a time. "That was from five in the morning until nine at night. All the daylight hours," he says. "That was as hard as I ever worked." As the second son in his family, Till knew he wasn't going to inherit any land, so he set his sights on science. He thought, "What I should try and do is the hardest science I can find, to see if I could do it. I thought, if I can't do it well I might as well go back and farm. No point in doing it badly." He immersed himself in the world of physics, earning a B.Sc. and master's degree from the University of Saskatchewan. His good marks won him a fellowship that sent him to Yale to do his Ph.D. in biophysics, which he completed in 1956.

Despite being offered a position as assistant professor at Yale, Till returned to Canada to set up a research laboratory in the Ontario Cancer Institute. "Here was this empty set of rooms with nothing in them. Just bare walls!" he exclaims. "And our job was to get something good going in research." He says the collaboration between disciplines was very exciting, with scientists and physicians getting together to discuss their work. McCulloch was interested in bone marrow transplantation; his goal was to treat leukemia with complete body radiation, and restore the normal blood cell function. He and Till started working together because McCulloch needed someone to measure exact dosages of radiation for mice, since overdosing would render the experiments useless. One of the areas Till specialized in was dosimetry, the measurement of doses of radiation.

"From the beginning . . . one of the best parts of our collaboration was that we didn't say, This is my territory, this is your territory," recalls Till, who thinks the open relationship was established because of an error he made. Observing the first irradiated mice, Till felt the survival rate seemed abnormal. He remembers wondering if he had administered incorrect dosages. He went back to his records and, sure enough, there was an error. He timidly told McCulloch what had happened. "Here was my expertise in question. I was very embarrassed at the time," he says, explaining that scientists and physicians "pride themselves in how accurately they do things. . . . But what it did, I think, was establish that we didn't pretend to be perfect. . . . And that meant we began to talk about all aspects of what we were doing, and wherever the idea came from, we didn't care whose idea it was."

Their historic discovery actually happened by accident. McCulloch transplanted irradiated marrow and normal marrow into mice to determine if the radiation affected growth properties. "We wanted to see whether radiation had any effects other than preventing proliferation [expansion of a cell population]," he says. "We wanted to see whether it affected the way the cells grew, whether the differentiation process—by which cells acquire a functional capacity—was affected by radiation. And the answer was no." But the experiment had some unforeseen and exciting results. One Sunday afternoon, McCulloch went to the institute to take some mouse samples. He looked at the mice's spleens, and found that lumpy areas had formed in the organs. The lumps were piles of cells that had divided and differentiated into several types of cells. Looking more closely, he realized that he could count the lumps. "What was novel was the thought that what you were seeing in the microscope was end-stage stuff," he says. "And all the really important things had happened much earlier." He also observed that there was a linear relationship between the number of injected marrow cells—irradiated or not—and the number of lumps.

"That's the form of the relationship that you obtain when a single entity causes an event," he says. "So that if you inject one and you get one, if you inject two and you get two, then if you plot [the ratio], it will be a straight line." The importance of a linear relationship, says McCulloch, "laid open the possibilities that these might be clones. In other words, we might be detecting the activity of a single cell." A clone being a line of cells that is genetically identical to one cell. McCulloch didn't run through the corridors with excitement. Rather, he went back to his office to plot the linear relationship. "I thought it was important," he says. "But I certainly didn't appreciate how important it was that first moment."

In the early 1960s, there was quite a controversy over the common origin of a lineage of different types of cells. Some scientists felt each lineage had its own specific origin, while others deduced that all lineages arose from one, single origin. McCulloch and Till proved that the latter view was correct. They discovered that the clones were essentially units of regulation. "The first thing we found out was that the number of clones that we were isolating were responsible for recovery after radiation," says McCulloch. "That particular kind of cell was the reason the heavily irradiated animals survived." They had in fact identified stem cells.

The difficulty came in trying to identify these stem cells—or colony-forming units (CFUs), as they were first called. It wasn't as if you could gaze into a microscope and pinpoint these rare cells; there was nothing particularly distinctive about them. "Think of a crowd of people attending a game, and one of them is a genius," McCulloch says. "They're not wearing different clothes or behaving differently. They're watching the game like everybody else." Till says they used to joke about what their cells looked like. "People tried to find stem cells. They thought they would look different. Have an S tattooed on their skin, or something. People would ask us, What do stem cells look like? And so I got so sick [of it] . . . I said purple. Royal purple. It became an in-joke with us." Stem cells, then, must be identified by their actions. "The trouble is that there is a philosophical problem," says McCulloch. "Because once you have it [are observing the cell], it changes. If it's alive, it's going to do something. It's the old uncertainty principle. The fact of studying it changes it. You're manipulating it."

What scientists found was that these "royal purple" cells didn't "breed true," which makes studying their genetics virtually impossible. The study of a genotype, or genes, usually depends on identifying the phenotype, or proteins. McCulloch offers some examples. "If you have a cake and you cut a piece—you've enjoyed it and think you'll have another piece—it breeds true if it tastes like the first one. In cruder terms, you take a family. If the father and mother are black, their children are likely to be black. They both breed true." The experiment suddenly became very perplexing. McCulloch remembers dissecting the stem cells from a single colony and injecting them back into the animal. The cells that were returned to the mouse did not reappear in their initial form. "So that if you took this one colony," he says, "you might get nothing, you might get many [kinds of cells]. It did not breed true."

The scientists witnessed some cells producing only one type of blood cell while another developed into several types, all in an apparently unregulated, haphazard way that made things even murkier and more mysterious. One day Till was browsing through the University of Toronto's physics library when he came across a book on cosmic ray showers. Niels Arley's *On the Theory of Stochastic Processes and Their Application to the Theory of Cosmic Radiation* described cosmic rays that produced a shower of particles because of a nuclear reaction between the cosmic rays and molecules within the atmosphere. The particles being produced differed in composition. This variation reminded Till of their own experiment. "The heterogeneity of composition of the cosmic ray showers," he remembers thinking, "is exactly the kind of heterogeneity that we're seeing with the progeny of these colony-forming units."

With his book, Arley, a theoretical physicist, had introduced Till to the stochastic theory, which deals with things happening by chance. Although the possibilities are fixed—either one thing can happen or another—the actual outcome is arbitrary. The theory proved to be very controversial when McCulloch and Till applied it to their work. "We don't like the notion of randomness," says McCulloch. "It's the basis of religion that there is some great being up there that's guiding us. [However] it is not incompatible that, if there is some divine thing, it happens to use stochastic methods." Life is an interconnected chain of random events. Think of digging your hand into a bowl of jellybeans, says McCulloch. You have just as great a chance of grabbing a red one as a green one. Or think of the whole mating process; the person someone decides to marry is chosen stochastically, he says. The couple happened upon each other by chance. Even today, some scientists disagree with the pair's stochastic conclusions. "As a species we tend to be stuck up, I think," concludes McCulloch. "We think we make a difference. We make decisions. The stochastic idea is that much of it happens by chance."

There was a computer at the University of Toronto, "a big, old clanking thing with lights flashing," says Till, that used a series of punch cards to run computer simulations of processes. The punch cards were fed into the computer, which translated their information into a binary code of zeros and ones representing the birth and death process, the option of a cell's living or dying. The stem cells could either "die" and differentiate to become a completely different type of cell, or "live" and continue dividing. In theory, though, each cell had both capabilities. "It's the so-called Monte Carlo approach," says Till. "You gamble. You throw the dice." With the two probabilities defined, the physicist let the computer make choices at random. They found that the distribution of the new stem cells per colony was as heterogeneous as the results spat out by the computer. While this model couldn't predict the distribution of stem cells per colony, it could predict the stem cells' characteristics. The stochastic process worked. Life was in fact random.

"Darwinian evolution is based on randomness, and geneticists are thoroughly familiar with random mutations," says Till, adding that such chaos can be quite disturbing. "Why [resort] to randomness unless you have to?" he asks, adding that as a scientist you would "like to think that there are compounds, molecules that you can put into a cell and the cell will behave differently, or become a stem cell when it isn't." Most scientists accepted McCulloch and Till's stochastic theory on stem cells. But even if everyone could agree on what a stem cell was capable of, and why, what good was this knowledge if it couldn't be controlled? For years these cells continued to slip through the grasp of scientists. McCulloch and Till could say for certain that stem cells existed. But it would take the experiments of an old Montana boy to pin these deceptive cells down and isolate them.

---

Dr. Irving Weissman remembers reading McCulloch and Till's remarkable paper on stem cells as a first-year medical student at Stanford. For him, it was like a light bulb suddenly went on. According to Weissman, McCulloch and Till didn't simply contribute to the field of stem cells, they launched it. Weissman's jovial face is framed by a well-manicured beard. He has lost most of his hair, yet is distinctly handsome, dressed in a sports coat and exuding a relaxed air of confidence. Forever in the public arena, debating the merits of stem cell research, Weissman is well-spoken and passionate in his refusal to let politics get in the way of science. When he is around friends, Weissman is famous for indulging in the pleasures of life. "He enjoys the food he eats. He enjoys the friends he has. He enjoys to travel," says Weissman's old friend Max Cooper. "He realizes that he's not going to live forever, and he enjoys life."[8] According to Cooper, Weissman is known for his extensive wine cellar and lavish dinner parties, where guests feast on rich foods and stories of his travels. He is an avid fly fisherman, and returns home to Great Falls, Montana, to cast his line any chance he can. Fly fishermen are hooked on the act of catching the fish.

The prize is the challenge, not the fish, which are thrown back into the water. Cooper remembers floating down the Bitter Root Valley with his friend on blow-up cushions, searching for a bite. Weissman, he says, was much luckier than he.

When it comes to science, Weissman has always had the same zealousness. In grade four, the Montana boy was mesmerized by Paul de Kruif's *Microbe Hunters,* which profiled history's great scientists—including Louis Pasteur, who developed a vaccine for rabies, and Paul Ehrlich, who discovered that the lethal substance arsenic could be used to heal. According to Ehrlich, "We must learn to shoot microbes with magic bullets."[9] The young Weissman decided to make this quest his own. In high school, he worked in a laboratory and published two papers before he turned 18. After attending Dartmouth College and Montana State University, he entered Stanford Medical School in 1960. Upon graduation, he shuddered at the thought of hospital board meetings and administrative duties, and vowed to focus on research rather than practise medicine. He joined the Stanford faculty, infecting students with his love of science. You can see it in the twinkle of his eye, his smile, and his boundless energy, says Cooper.

In the early days of his research, Weissman was consumed by the tragedies of Nagasaki and Hiroshima, and how radiation had wiped out survivors' blood cells, often handing them a death sentence. He studied how the immune system wards off cancer, examining the connection between blood cells, cancer, and radiation. Bone marrow transplants were being used to stop the fatal effects of radiation. The transplanted marrow renewed the blood that had been poisoned, so Weissman knew that the marrow housed cells with regenerative capabilities. Years had passed since McCulloch and Till had identified blood stem cells, but they remained fresh in his mind. "I knew those experiments," says Weissman. "I knew them cold."[10] He was determined to isolate these cells.

Weissman accomplished this by sifting through mouse blood to remove unwanted cells, as if he had a series of strainers.[11] To do this he used monoclonal antibodies that could identify cell surface proteins that are not found on stem cells. One type of antibody, for example, would pluck out mature bone marrow cells, which have a certain type of surface protein. With each round of straining, he removed a different type of cell—the T-cells, the B-cells, the red blood cells, the white blood cells—until he was left with what he called Lin-, or lineage minus, since the antibodies had removed all other cell lineages.

When his blood straining was complete, 90 percent of the bone marrow had been discarded, leaving Weissman with 10 percent that showed "stem cell activity." To home in on his target even more, Weissman used two other monoclonal antibodies that seek out surface proteins of stem cells. He was left with 0.5 percent of the cells, and found that only 30 of the cells were needed to regenerate the blood system of an irradiated mouse.[12] "Weissman's remains the most extensive purification that's been achieved," says McCulloch. "I consider his work to be seminal. The application to man was just a step beyond what he did in the mouse."[13] It was 1988, and Weissman had finally isolated blood stem cells.

For the next decade, the race was on to separate and capture embryonic stem cells in culture. Scientists could then pause the biological clocks of stem cells and create millions of hollow little amnesiacs that had lost their way. They would live in the moment, with no memory or direction, until they were given a purpose and a road map that would lead them to become muscle, blood, even brain or bone. Scientists like McCulloch, Till, and Weissman understood that embryonic stem cells could reshape humanity. Hematopoietic stem cells had transformed the treatment of blood diseases. But by harvesting the raw material of life, biologists would propel a revolution in medicine. For years, the hunt continued. Then, in 1998, two scientists in separate laboratories announced that the race was over.

---

John Gearhart spent the first six years of his life exploring his family farm in Western Pennsylvania's Allegheny Mountains. This freedom was taken from him when his father, a coal miner, died, and he was sent to an orphanage for boys. Gerard College had a 12-foot-high wall, and for a decade Gearhart was surrounded by concrete. "We were only outside these gates for brief periods of time so it was really like a little, you know, sequestered—although there were a thousand kids in there," says Gearhart. "It was a very isolated kind of a thing." His new home, he says, could have been mistaken for the Eastern State Penitentiary, just up the street.[14]

Life at the orphanage was regimented. By 5:30 a.m., students were reviewing their lessons in the study hall, where they also spent their evenings. To this day Gearhart continues to rise at daybreak to work and read. In high school he became interested in agriculture and how science could enhance crops. A counsellor announced that he would never amount to anything, and told him not to waste his money on university. But Gearhart didn't listen. He used his dad's Social Security savings and enrolled in Penn State University, where he earned a B.Sc. in 1964. For his Ph.D. at Cornell University, he studied the fruit fly *Drosophilia melanogaster*. Then he joined the faculty at Johns Hopkins University.

Always fascinated by the way things worked, at Johns Hopkins Gearhart studied all forms of life, from plants to animals, finally focusing on humans. Many mouse genes are similar to those found in humans, making the shift in research relatively effortless. Gearhart became particularly interested in Down syndrome, a chromosomal disorder. What if too many genes in a person's genome caused this, he wondered. How did someone's balance of genes affect his or her development? He soon noticed that the biological glitches associated with the syndrome seemed to happen extremely early in life. This was the early 1990s, and Gearhart thought of the embryonic stem cells that had recently been isolated in mice. If the same could be done for humans, thought the biologist, controlled cells could be cultured to form any of the body's tissues. He took the same cells and cultured them with an extra chromosome, to see how the tissue was affected by the extra sets of genes. He was then able to determine how the chromosomal disorder affected the body's different cells.

In 1998, Gearhart made scientific history. After many unsuccessful attempts, he and his team of Johns Hopkins researchers isolated human germ (reproductive) cells from the developing gonads of five- to nine-week fetuses that had already been aborted. He encouraged these cells to continue multiplying, and was able to keep them dividing in culture for more than seven months without differentiating. They had found the biological Holy Grail: a self-replenishing supply of cells. The body could rejuvenate itself! The cells that trigger disease could be replaced, and health restored. Life would never be the same. "I think for most scientists the fun is in the journey. . . . Of being in the laboratory, of knowing that you can do whatever experiment you want to do," says Gearhart. "It has to come from you, and you have to think about what the problem—what the questions are, how you plan that experiment, how you execute it, what controls you use. You know, interpretation. This to me is the excitement."[15]

Gearhart shared the glory with a painfully shy developmental biologist at the University of Wisconsin–Madison. James Thomson would rather be hang-gliding or sea kayaking than talking to the media. He also flies vintage model planes from the 1930s and 1940s. A rocket-scientist uncle at NASA made the field of biology seem exciting and adventurous, and Thomson quickly discovered this fascinating world. He earned Ph.D.s in both veterinary science and molecular biology. It was Thomson who first isolated embryonic stem cells from the inner cell mass of blastocysts. While Gearhart used aborted fetuses, Thomson used donated embryos that had been stored at fertility clinics and were no longer of use. Typically, many embryos are stored to help impregnate a woman. Once she becomes pregnant, the excess embryos are discarded. But for Thomson, a non-practising Congregationalist, the decision to use embryos for research was morally and ethically challenging. He read everything he could on the subject, and sought advice from bioethicists including the University of Wisconsin's R. Alta Charo and Norman Frost. Finally, Thomson concluded that it was better to advance science and one day help humanity by experimenting with discarded embryos than to leave them to be tossed in the trash.[16]

For Thomson's famous experiment, embryos were given a special "nutrient broth" that encouraged them to grow to the fifth day of development. At this point, Thomson and his team of scientists could extract stem cells from the blastocyst's inner layer of cells. Creating an ideal biochemical environment, Thomson encouraged the stem cells to multiply industriously without differentiating. The enzyme telomerase, which is found in all cells, is what keeps these cells from aging and prevents them from dying like typical cells that have limited lifespans. Then, magic. When Thomson eliminated the biochemical signals, the cells were left to their own devices and began changing into specific cell types, like bone marrow, heart muscle, or kidney cells. The billions of stem cells were ready to be injected back into the body to repair weakened tissue and organs.[17] Suddenly, the body became its own tool kit for repair and regeneration.

Overnight, the son of an accountant and secretary was pulled from his isolated little world. His image was splashed across TV screens and magazine covers. Back in 2001, Thomson said the press made it difficult to get any work done at all. Today, he agrees to only a select few print interviews, and refuses television appearances. His home, which he shares with his neurobiologist wife and two kids, doesn't even have a television. "Jamie is very self-effacing—the last person in a lineup [you'd] point to as one of the biggest superstars of science in the 21st century," says bioethicist Norman Frost.[18]

Yet that is exactly what Thomson is. Along with Gearhart, he catapulted medicine into another era. These scientists deconstructed biology, uncovering the secrets of those mystical stem cells. These magic seeds of life may one day create blankets of skin for burn victims, bone for broken or osteoporotic limbs, or pumping muscle for heart attack survivors. The door to the clinic is now open.

---

People travel thousands of miles to stand beneath the cloudless cerulean sky of the desert and inhale its restorative heat. The dry air is as intoxicating and penetrating as a sauna, loosening muscles and slowing movement. The absolute stillness inspires self-reflection and spiritual awakening. The blinding sun invigorates the body, while the heavy ink-black night cools the soul. People come here for personal regeneration. Some lift their spirits with Native American herbal treatments. Others pamper their aching bones by escaping the ravages of winter. Dr. Samuel Cohen came to Arizona in search of a new heart.

The heart's job is to pump oxygenated blood to the rest of the body. It's not easy work—even at rest, the contractions of the heart muscle are pumping about five litres of blood per minute—so the heart muscle itself needs to be nourished by a steady supply of oxygenated blood. The heart's own blood supply is delivered by the coronary arteries. In a heart attack, part of a coronary artery becomes blocked, and the area of the heart muscle downstream from the blockage suffers a lack of oxygen. Unless the oxygen supply is restored very quickly, that part of the heart's muscle will be destroyed. Some people die suddenly from a massive heart attack; others have one minor attack after another, losing some heart function each time. The more muscle lost, the less efficient the heart is at pumping oxygenated blood through the body, which leads in turn to weakness and fatigue. As well, the remaining heart muscle becomes overworked, trying to compensate for the muscle that is no longer doing its job.

After three heart attacks, the 61-year-old Cohen made his pilgrimage to the Arizona Heart Institute in Phoenix. It seems fitting that a centre to repair and renew weakened hearts has found a home in a place long revered for its healing powers. The sand-coloured buildings blend in with the desert, while heavy, leaning palms and manicured grounds offer an oasis of green. Inside, the waiting room is as inviting as a hotel lobby, with plush carpeting, paintings, and sofas you can sink into. The cardiac conditioning facility resembles a hotel gym, with rows of shiny new machines, mirrors, and huge windows. The admittance desk could be mistaken for a check-in counter, with bowls of fruit and lush plants. This is the look of regenerative medicine: the next frontier of health care.

Dr. Cohen is here to participate in a Phase I clinical trial to test the safety of repairing damaged hearts like his own with myoblast cell transplantation. This type of transplantation uses undifferentiated myoblast cells, which develop into muscle cells. The myoblasts are injected into areas of the heart that have been destroyed by heart attacks to regenerate them. Scar tissue has only 20 percent as much blood flow as that through the tissue of a normal heart. The myoblasts replace the scar tissue with muscle tissue, which regenerates the heart and the patient's blood flow. Cohen arrived here with a temporary pump in his heart to keep him alive. His heart had slowed to 28 percent capacity, making him a semi-invalid. Walking was arduous, and basketball games with his son had been forgone. "You lose the feeling of invincibility when you have heart problems," he says.[19]

Dr. Nabil Dib, an interventional cardiologist and chief of Cardiovascular Surgery at the Arizona Heart Institute, is the trial's principal investigator. His goal is to determine whether myoblast cells injected into the dead tissue of damaged hearts can regenerate the scarred area. There are three types of muscle in the body: skeletal muscle found in arms and legs; smooth muscle, which makes up the gastrointestinal tract and is responsible for making stomachs growl; and cardiac muscle, which pumps blood 24 hours a day, seven days a week.

Unlike embryonic stem cells, myoblasts are destined to become muscle. But like specialized stem cells, they can morph into different types of muscle cells and multiply by the millions. Say you injure your bicep. The myoblasts hiding within this muscle will switch themselves on to churn out new myoblasts and repair the wounded arm. Yet these cells are limited in what they can do. If you suffer a heart attack that destroys cardiac cells, these cells cannot repair themselves. This is where Dib comes in. He has figured out a way to regenerate cardiac muscle.

This cardiologist has always thought with his heart. "To me life is emotion. And emotions attract me a lot," he says, adding that the dynamics of the heart drew him to cardiology.[20] Dib grew up in a small town in Syria without electricity. He read by candlelight and drank water from a lake. At 12 years old, he left his family for a bustling city to attend school. At 18, he won a scholarship to go to Damascus University School of Medicine, where pre-med and medical school were combined in one six-year degree. He interned in New York and Boston before doing an internal medicine residency at Boston's Tufts University School of Medicine. In 1999, he joined the Arizona Heart Institute to repair the hearts of patients like Cohen.

Weeks before his surgery, Cohen enters the operating room for another procedure. Dib must harvest Cohen's myoblast cells for the transplantation. By using the patient's own cells he avoids the risk of rejection, and the possibility of transmitting an infectious disease. Cohen is given local anesthetic, and Dib scoops out a muscle biopsy sample from Cohen's thigh that weighs between two and five grams. The sample is taken to an outside laboratory for a process of mechanical and chemical dissociation that isolates the myoblast cells from the other types of muscle cells. These myoblasts are then encouraged to multiply for the next month; this is the same type of self-renewal found in cultured stem cells. When there are enough cells for the surgery, Cohen returns to the hospital.

Since the purpose of a Phase I trial is to test safety rather than efficacy, patients cannot receive the myoblast transplantation on its own. They also require bypass surgery. This means that they need to have a coronary artery that will accept the bypass, and there must be enough tissue in the heart to work with. Cohen is one of 27 such participants. He is prepped for the standard open-heart procedure, which Dib performs. At the same time, the myoblast cells are warmed for transplantation. The bypass takes four hours, after which Dib begins the experimental cell therapy. Cohen's chest remains open the entire time. Dib aligns the heart with an electromechanical mapping system, which allows him to find the most damaged regions of the heart, and map the targeted areas for injection.

Ten million cells are poured into a syringe that injects them into the dead tissue. This takes 15 seconds, and Cohen receives 30 such injections in less than 10 minutes. Three hundred million myoblasts dive into the pit of scar tissue and are left to regenerate the paralysed muscle in Cohen's heart. When heart muscle is damaged, it forgets how to contract. The damaged areas of the heart will relearn the process of contraction. With both procedures complete, Dib closes up the chest of his patient, who is kept at the institute for observation for five days. Then Cohen is free to go home.

Six months after surgery, Cohen's PET scan shows that his heart function has improved. He is at 40 percent capacity compared to the 28 percent capacity he had going into the procedure. While bypass surgery

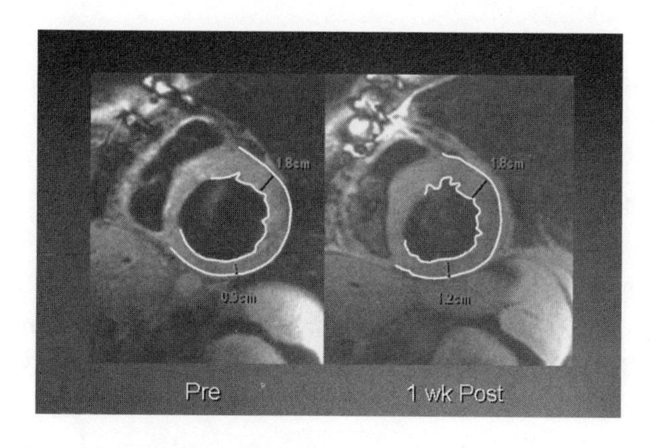


**Samuel Cohen's heart wall thickens from 0.9 cm to 1.2 cm at the site of heart attack one week post surgery.** —Arizona Heart Institute

alone would have helped to strengthen Cohen's heart, Dib says the cell therapy offers more promising results. Among the study's 27 participants, he says, "We found that the transplanted myoblasts survived and thrived. Areas damaged by heart attack and cardiovascular disease showed evidence of repair and viability."[21] Dib explains that the study successfully assessed the safety and feasibility of the procedure. There were no adverse events to report. Now that myoblast transplantation can regenerate scar tissue without increasing the risk of death, a Phase II trial, to test efficacy, can begin in the near future. "This is really leading-edge technology. This is where medicine will be going in the future," says Dr. Edward Diethrich, founder of the Arizona Heart Institute. Cell therapy will soon change the way heart disease is treated.

Diethrich and Dib are currently working on a less invasive procedure that could one day render open-heart surgery obsolete. An electromagnetical mapping system creates a three-dimensional, colour-coded image of the heart that also measures the voltage in the left ventricle. (The voltage is typically low in heart-attack patients.) The damaged area is highlighted and targeted. A catheter is inserted in the patient's groin and fed through the femoral artery, winding its way up to the heart. Having tested the procedure on pigs, Dib says the catheter can inject cells within one or two millimetres of accuracy. Once the institute receives FDA approval, it will begin human trials to test the safety of the procedure.

Not only will this minimize the invasiveness of surgery—easing recovery and shortening hospital stays—it will offer hope for patients too weak or elderly to have open-heart surgery, and those who are in need of a new heart but cannot find a suitable donor. This can't come too soon for the 12,000 Canadians who suffer heart attacks every year, and the 500,000 Americans who die annually from heart disease. Heart disease is the second leading cause of death in Canada. Two million Americans are admitted to hospital with congestive heart failure every year, says Dib, adding that the chance for survival is far bleaker than with cancer.

Teaching the heart to regenerate itself is Diethrich's vision for the Arizona Heart Institute. Since he launched the institute in 1971, he advanced heart surgery and therapy. Over the past three decades he has perfected innovative procedures that are becoming less and less invasive. His experiments with regenerative therapies have placed him in the forefront of a whole new approach to the treatment of cardiac disease.

Outside the operating room, the dashing doctor could be mistaken for a movie star. He is tall and slim, with a year-round tan and sparkling white teeth. But the son of a teacher and a nurse was destined for medicine. As a young boy growing up in Toledo, Ohio, Diethrich was already playing doctor. He would snip open his teddy bears, pull out the cotton, and methodically replant it back into the stuffed animals. He fondly remembers his mother in her peaked white cap, nurse's pin, and uniform, and has the nurses at the Arizona Heart Institute uphold this image. Before even graduating from high school, Diethrich was already assisting a local doctor, suctioning for tonsillectomies and performing his first vasectomy.

While studying medicine at the University of Michigan, Diethrich scrubbed and prepped patients in the operating room at St. Joseph's Mercy Hospital, where he stayed to do his surgical residency. As an intern, he devised a new tool for heart surgeons to split the breast bone and open a patient's chest for an operation. Before this, surgeons used a crude knife-and-mallet system to chisel open the chest. Diethrich bought a sabre saw and modified it to suit the procedure.

It was the 1960s, and Diethrich was fighting his way through a two-year thoracic and cardiovascular surgery residency at the Baylor College of Medicine, in Houston, under the guidance of renowned heart surgeon Dr. Michael De Bakey. New equipment was expanding the field of cardiac surgery. Transplants were becoming routine, but there weren't enough donors to go around. So a team of doctors that included Diethrich performed a multiple-organ transplant, taking the heart, one lung, and two kidneys from a single donor and treating four patients. Medical history was made. But for Diethrich, this was just the beginning.

In the early 1970s, Diethrich decided to break out on his own. He headed for the desert, where he launched the Arizona Heart Institute. In 1979 the institute opened the first outpatient cardiac catheterization clinic. Doctors were dumbfounded. "It just bucked everybody's conservative, frozen mindset," says Dr. Robert Strumpf, the institute's director of Interventional Cardiology.[22] The Arizona Heart Institute, he adds, does not accept "progress by committee." Diethrich has never allowed other people's opinions to dictate his career: "It isn't that I want to buck the establishment," he says. "It's that I don't really care what they say or what they think. . . ."[23]

This is good news for patients like Samuel Cohen, who don't have time to wait for the mainstream to approve. Myoblast cell transplantation has given Cohen's heart new life. He feels stronger. In fact, he says he hasn't felt this strong in years. He now exercises for half an hour every day, and climbs onto the treadmill a few times a week. And, more important, he is back on the court with his son, playing basketball.

## Chapter Five

# The Body Builders

## Tissue Engineers Reconstruct Broken Bones and Ailing Parts

Outside Molly Shoichet's office window, men in hard hats are slowly erecting the skeleton of a future building. For months they remained on the ground, creating a solid foundation to bear the weight of this massive structure and everything it will house. Over time, the construction workers made a slow ascent, scaling an intricate wall of scaffolding. Now cranes hoist heavy slabs of cement. Staccato drilling shakes the sidewalk below. In the morning, light filters through the space between steel and cement, casting shadows of men hard at work. If Shoichet could spare a moment to look out her window, she would come face to face with these men, who are now working at eye level with her.

Shoichet is also in the business of construction—or, to put it more precisely, reconstruction. In her University of Toronto laboratory, the professor of chemical engineering and chemistry constructs cellular scaffolds to regenerate damaged spines. These tubular constructs are made of the soft plastic used in soft contact lenses, which can be moulded into virtually any size or shape. They are also biodegradable, which means they break down inside the body when they are no longer of use. The idea is to reconnect severed spinal cords by implanting these scaffolds in the area of dead tissue where an injury has occurred.

Imagine an electronic cable that has snapped in half. One piece has become two, silencing electronic messages. To reopen the line of communication, something must bridge that gaping space between the two ends of broken cable. Transplanted into the spine, Shoichet's polymer scaffolds encourage nerves to crawl towards each other across this cellular bridge and begin talking again. Once these nerves are reconnected, they can tell a paralysed body how to move, and how it is feeling. "Our bodies have lots of internal plumbing," says Shoichet, "which polymers can be used to mimic."[1]

This is all happening in an old, unassuming three-storey stone building on the University of Toronto campus, at the corner of College and University Avenue. Climb the creaking wooden stairs of the Mining Building up to the third floor, follow the hallway of exposed brick and bulletin boards exhibiting the latest published papers, and you will come to a locked door. If you peer through a small window, you will see students in lab coats bent over their beakers and test tubes, laboriously measuring out the compounds for their latest experiments. It is an exacting profession: if these gloved hands squeeze one too many or one too few drops into the solution, the day's work could be lost. But if it all goes according to plan, the magic of regenerative science takes hold. Things are witnessed for the first time by wide-eyed, expectant students. As possibilities are ruled out, the students narrow the field, closing in on the answers they seek. Experimenting with different types of constructs, Shoichet and her team of budding scientists intend to find the ideal chemical, mechanical, and physical properties needed to reawaken a body that has been silenced or immobilized by trauma.

Shoichet is a petite, soft-spoken, 40-year-old woman with chocolate-brown hair and delicate features. The scientist-cum-entrepreneur often dresses in suits so she can dash off to board meetings for matREGEN, a tissue repair company that she co-founded. In her lab, Shoichet tries to forget about money and projections, and focuses on a group of lab rats she is teaching to walk again. After countless months of trial and error, her lab rats have managed to take a few celebrated, clumsy steps. At the beginning of the experiment, the animals had the movement capabilities of a paraplegic: they were unable to control the lower half of their bodies. A surgeon implanted Shoichet's scaffolds into the peripheral nerves or spinal cords of the rats, and the animals were left to recover. Miraculously, the damaged nerves began to regenerate. Rather than dragging their immobile hind legs around the cage, the rats were able to take their first awkward steps. Shoichet's scaffolds have given these rats new legs to stand on.

While experiments in rats can offer clues to the mysteries behind spinal cord injury, no one really knows what will happen once these scaffolds are implanted into human spines. Our spinal cord does in fact harbour the ability to recover after injury, with limited results, yet our regenerative capabilities are different from those of mice, which are commonly used in laboratory experiments. When we experience trauma to the spine, a cavity may be left at the site of the injury. In mice, the injured area is naturally plugged with a tissue matrix that conveniently brings the severed nerve endings closer together.[2]

The brain and the spinal cord constitute the central nervous system. The spinal cord is a pipeline of nerves—about 18 inches long in adult humans—that runs from the base of the skull to the waist, carrying messages between the brain and the rest of the body. The motor and sensory nerves in the rest of the body, connecting to the spinal cord, make up the peripheral nervous system. Sensory pathways warn us if we touch something scalding or frigid, or if we are in pain. Nerve pathways tell us what our organs are feeling, and when something feels wrong. Protecting the spinal cord are vertebrae, halos of bone that we know as the spinal column, or backbone. All in all, our spinal pipeline consists of 33 vertebrae, 31 pairs of nerves, 40 muscles, and countless tendons and ligaments.

Spinal cord injury can occur for many reasons. Sudden trauma from a car accident, a serious fall, or a gunshot can damage the spinal cord, as can the progression of a disease such as polio or spina bifida. The degree of damage to the spinal cord depends partly on the type of injury: complete or incomplete. After a complete injury, there is no function below the trauma site. An incomplete injury allows for some movement or sensation below the point of damage. Those with incomplete spinal cord injuries can often move one limb more than another, or experience sensations in limbs that they are incapable of controlling. Loss of function and sensation does not necessarily mean that the spinal cord has been severed. A wounded but intact spinal cord could result in similar immobility. After an injury, the spinal cord often swells. In the days and weeks following, patients must wait to see if they will regain any movement or sensation. This can occur as late as 18 months after the trauma.

The location of the injury is also a crucial factor. Vertebrae are identified according to where they are found in the body. The neck houses eight cervical vertebrae—C-1 through C-8—that enable it to flex, bend, or twist. Damage to these vertebrae can result in quadriplegia, paralysis of both arms and both legs. If an injury occurs above the C-4 vertebra, the person may lose the ability to breathe independently. An injury above the C-5 level may suppress hand and wrist mobility. At chest level there are 12 thoracic vertebrae, T-1 through T-12, responsible for rotating and bending movements in the spine. Injuries to the thoracic vertebrae can result in paraplegia, in which movement of the lower body is muted. Trauma between T-9 and T-12 can affect control of the trunk and balance while sitting. People with injuries between T-1 and T-8 may have hand movement but inadequate lower body control. The lower back houses the lumbar vertebrae, L-1 through L-5. Damage to this region can affect hip and leg mobility.

The autonomic nervous system helps direct the bodily functions over which we have little or no control. It has two components—sympathetic and parasympathetic—which can likewise be affected by spinal cord damage. One function of the sympathetic nerves is monitoring blood pressure and heart rate, controlling the blood flow to the heart; spinal cord injury can cause these nerves to become overactive. Parasympathetic nerves are responsible for, among other things, dilating blood vessels to slow down the heart. Even after a spinal injury, these nerves labour to control blood pressure.

There is currently no cure for the approximately 36,000 Canadians and 450,000 Americans who live with spinal cord injury. Every year there are about 10,000 new cases in the United States, and 1,050 in Canada. The leading cause of these injuries in both countries is motor vehicle accidents. In the U.S., the second leading cause is violent crimes, which account for 36 percent of all cases, followed by 21.2 percent from sudden falls. In Canada, falls (including industrial accidents) are the second cause of spinal cord injury, at 17.7 percent. Other causes include diving, other sports injuries, and medical conditions. If spinal cord victims can survive the first day after their injury, there is an 85 percent chance that they will survive the next decade. The most common complication is respiratory disease, including pneumonia, which is the leading cause of death in the 15 years following the trauma for all ages and races, and both sexes.[3]

For decades, the best possible treatment for spinal injury has been an autograft, in which a surgeon removes tissue from one area in the patient and transfers it to the site of the injury. While there is no fear of rejection, because the patient's own tissue is used, the procedure causes a second injury that is often more painful than the one being treated. Scooping out bone cells from the hip and transplanting them into the damaged spine may promote spinal fusion, but it also means killing healthy tissue.

A fundamental goal of regenerative medicine is to discover the stimuli that will switch on the body's innate restorative capabilities, so that tissues and organs repair themselves. Scientists specializing in biomaterials (materials with properties that imitate or encourage biological processes) and biomedical engineering are creating biodegradable constructs that will help the restorative process get started. Cellular scaffolds promote what Shoichet calls "guided regeneration," encouraging tissue to grow in specific areas rather than sporadically.

---

As a little girl growing up on Toronto's tony Post Road, Shoichet thought she was destined for medicine. Even though her father, Irving, had set his sights on medical school before he chose to take over the family fuel-oil business, it was Shoichet's mother, Dorothy, who had the greatest impact on the youngster. The girl grew up watching her mom launch her own business, which expanded to two hundred employees. This gave the young Shoichet the drive and ambition to become a professional. Shoichet's brother, Brian, who is also a scientist, says it was drilled into his sister that she would pursue a career.[4] He says they were proud of their working mom, who was anything but an anomaly in their affluent neighbourhood of lawyer and doctor moms. Shoichet dreamed of becoming a physician while she was a student at the Toronto French

School, and carried this with her to university. As an undergraduate science student at MIT, she took all the pre-med courses. It was also during this time that she began her work in regeneration, experimenting with collagen tubing and trying to understand how damaged nerves might be repaired.

When Shoichet was accepted into graduate school as well as medical school, she didn't know which path to take. She had a pretty good grasp of what a doctor does, but wasn't as clear on the life of a scientist, so she deferred medical school and entered the MIT grad laboratory. Brian says she was more and more grabbed by the science, explaining, "once you get that hook in you, it's hard to let it go." Shoichet also found that she had the guts to pursue the perplexities of the human body. "There's a certain amount of intellectual courage to do science. You can get married to an idea, and it can fail. Ideas fail," says Brian. "And you have to be willing to take those sorts of psychological risks, and be ready to be wrong and sort of embrace that process. And Molly does that." Two years into her graduate degree, Shoichet found there were still many questions she wanted to tackle. Rather than practise the medicine that was already known, she would work on science that could one day advance medicine.

Brian says his sister possesses the intuition of a natural scientist: "It's the recognition of something out of whack, and 90 percent of the time the thing that's out of whack is a stupid artefact and you should ignore it. It's out of whack because you forgot to add something, the machine was having problems, or the moon was in Capricorn. But 10 percent of the time there's something there. And being able to distinguish that 90/10 thing . . . even when everyone else is saying that's stupid . . . following that through, that's the key. She has that indefinable thing, that 'right stuff' that Tom Wolfe was talking about."

Over the past few years, Shoichet's scaffolds have made her something of a celebrity in science circles. But she has her life in perspective. Sure, she is dedicated to her research and always keeps her eye on the prize, says her brother. But she is more committed to the cause she now represents than to her own image. "In her personal life she doesn't go around saying,

'I'm famous,'" adds Brian. "I don't think she's affected by it." And anyway, she may be famous in the world of science but in the outside world, he says with a chuckle, "she's not famous like Madonna is."

Molly Shoichet is more likely to boast about her two young boys, Emerson and Sebastian, than her walking rats. During a visit, she will most likely show you pictures of her kids, bragging like any proud mom. For hours you may sit with her in her paper-strewn office overlooking the construction site as she explains the challenges of nerve regeneration. Outside her office there seems to be a permanent queue of students waiting patiently to see her. But when you have her attention, she never appears hurried or stressed. Rather, she seems lost in her own world of cells and polymers, doodling pictures of spines and scaffolds to illustrate her ideas.

---

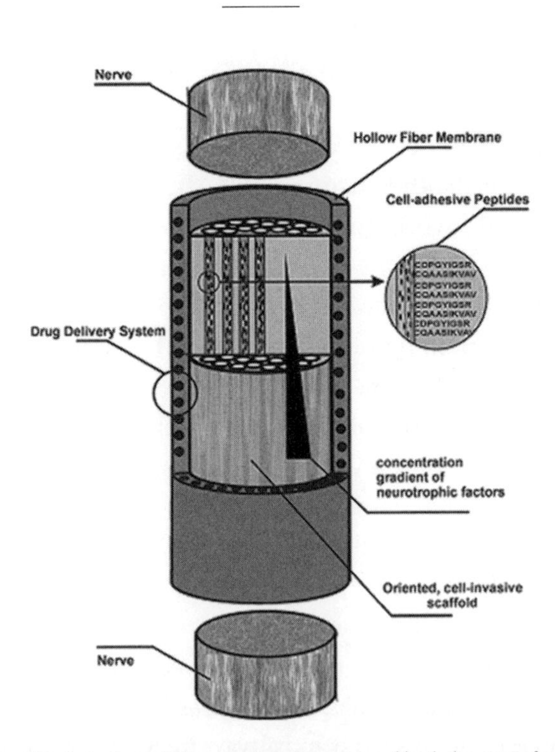


**Professor Molly Shoichet and her team propose that to repair the severed spinal cord, a nerve guidance channel provides a bridge for regeneration. This channel is filled with stimuli that guide regeneration across the injury site.**

—Molly Shoichet

Shoichet invented her own processing technology to create scaffolds. She begins with a horizontal tube with a vertical test tube rising up from either end. The main ingredient is measured into one test tube: a monomer, a single chemical compound that transforms into a polymer as it reacts with itself. A protein, a growth factor, is measured out into the other test tube, and blue dye is added so that the scaffold's rate of growth can be monitored.

Both solutions slide down their separate test tubes, meeting in the horizontal tube below. Slowly the blue solution crawls through the horizontal tubing, indicating that the materials are becoming fused. Inside the horizontal tube is the scaffold mould, which will become a dark shade of blue once the combined solution works its way into the plastic skeleton and becomes a polymer. The whole apparatus sits on a box that houses a magnet, which acts as a stir bar to keep the solution moving through the tubes. The monomer is left for an hour to transform into a usable polymer. Then the horizontal tube is carefully separated with a glass cutter, exposing a scaffold that looks like wet spaghetti. Since the scaffolds need to remain soft and flexible, they are stored in water.[5]

Shoichet must determine whether the polymer's chemical and mechanical properties are encouraging nerve development. To test her growth factor, a scaffold is sliced in half and placed in a mini Petri dish, where it sticks to the hard plastic bottom. Then the medium is poured in. Next, nerve cells are added, and float down through the solution to eventually suction themselves to the scaffold. For two days the cells are left to grow. Then they are stained and placed under a fluorescent microscope.

A nerve cell has a long axon and a crown of short dendrites—antenna-like fibres that extend from its body, their wispy tendrils outstretched to communicate with other nerve cells. During spinal cord injury, axons are damaged and fall into a coma-like state of non-responsiveness. It is the job of the growth factors to give these axons a wake-up call. To determine the direction in which these reawakened axons are growing, Shoichet's team examines which way the tendrils are pointed. If the growth factor is working optimally, the axons, which under the microscope look like tiny open hands frozen in time, will be reaching out to the

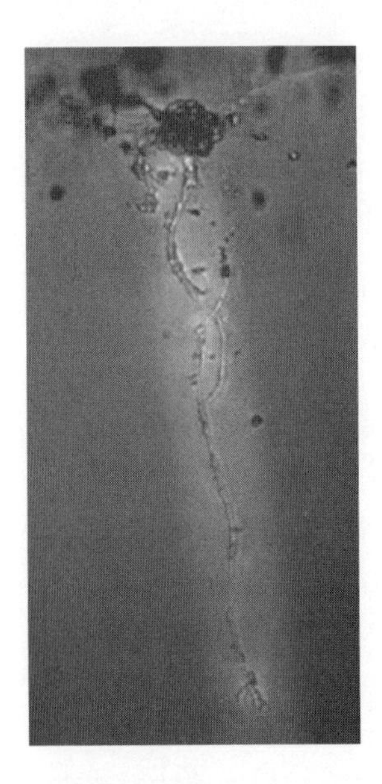

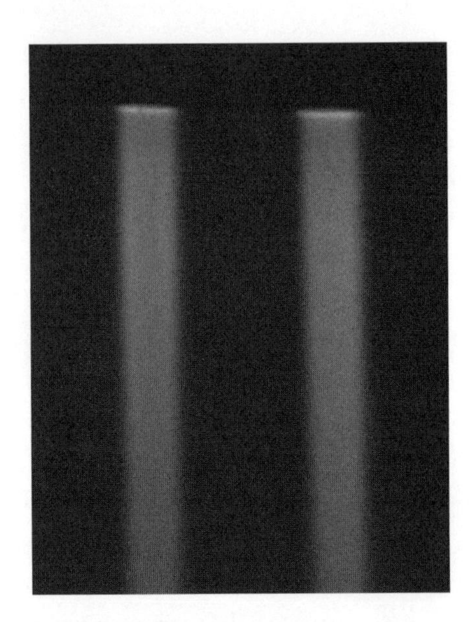

**(left) The biochemical pathways created with light guide nerve cell growth. (right) A new 3-D scaffold technology made by Professor Molly Shoichet and former PhD student Ying Luo uses light to create chemical pathways for guided regeneration.**
—Molly Shoichet

next protein along their path. When a particular scaffold appears to have the right properties to trigger nerve regeneration, its soft plastic outer layer imitates the surface of tissue. Yet it has the strength to withstand pressure at the transplantation site.

But Mother Nature is difficult to trick. Shoichet's goal was to regenerate axons by building a three-dimensional bridge for them to crawl across, so that they could reconnect with their neighbours. One of her biggest challenges was to guide surrounding nerves *through* the scaffold, not just along its surface. For years, scaffolds were restricted to two dimensions. Scientists couldn't seem to figure out how to guide these axons through the inside of a cellular tube without them clinging to the sides. The axons appeared to be drawn to the texture and feeling of a scaffold's hard outer layer, while the inside of the scaffold remained bare.

One of Shoichet's students, Ying Luo, solved this problem by creating an innovative three-dimensional structure. Most three-dimensional constructs have a maze of interconnected pores for cells to adhere to. By using laser technology, Luo created chemical channels that lured the cells inside the scaffold. The cells were then content to grow inside this three-dimensional network.

The body of each nerve cell in a damaged spine remains alive, so Shoichet must also deliver the appropriate stimuli to spur regeneration in each cell. In the central nervous system, she is trying to teach motor nerves, which control movement by sending messages from the brain to the rest of the body, to grow downwards. Sensory nerves, which carry messages from the body back to the brain, are encouraged to grow upwards.

Shoichet's rats have experienced improved mobility. The temporary scaffolds implanted in them are biodegradable, and disappear once they have triggered the regenerative process. But in order to get the rats walking normally again, Shoichet must devise a way to get nerve regeneration into the tissue beyond the site of injury. Even if regeneration over the implanted scaffold is perfected, she must still push this regeneration back into the surrounding tissue to completely connect it with its target. That target out in the periphery—whether a paw or a foot—must be taught how to move and learn how to feel again. It remains a long reach for axons to span the distance to an old connection.

There is a lot of work ahead before Shoichet's discoveries can be taken to the clinic. She says they need better results in rats before they can even consider moving to humans. But she is convinced that that day will come. It can take up to a year to perform a study and analyze the results to determine the level of progress. Tissue engineering has progressed by leaps and bounds over its short lifespan. Just give her time.

---

In 1995, American scientist Charles Vacanti shocked the world by growing an ear on the back of a hairless lab mouse. News programs ran the disturbing story, nauseating a public already sick over the dangers of biotechnology. Animal activists denounced the shock-and-awe experiment, calling it barbaric and futile. But Vacanti did have a point to all of this. He was looking for attention, and his freaky mouse certainly got it.

Six years earlier, Vacanti and his team at Massachusetts General Hospital had celebrated the successful growth of human cartilage in a laboratory. But the project was rejected hastily by a distinguished research journal. "The editors said, 'We believe you, but we can't see any practical implications,'" Vacanti remembers dumbfoundedly. "That was one of the most ignorant responses."[6] Undeterred, Vacanti questioned several plastic surgeons about body parts made of cartilage, asking which was the most difficult to repair. It was unanimous. The ear is the most vexing to mend and, surprisingly, there is always a great demand for new ears. Adults often lose an ear during car accidents. Some children are born without ears, while others have them bitten off in schoolyard fights. Often, ears become so mangled in accidents that they cannot be sewn back on.[7] Vacanti was reassured that there was a need for his research. Now all he had to do was prove that it could work.

In his laboratory at the University of Massachusetts, Vacanti, along with chemical engineering assistant professor Linda Griffith-Cima, constructed a scaffold that would mimic the shape of a human ear. The scaffold was made of porous polyester fabric that would biodegrade in the body over time. They chose a mouse as their host, and nicknamed it Auriculosaurus. The mouse had been bred to have a deficient immune system so that it would not reject the implanted tissue. Human cartilage cells were seeded along the ear-shaped scaffold, which was then transplanted onto the mouse's hairless back.

The animal would continue to grow its own cells, which would thread through the scaffold's fibre, eventually taking over the function of the support structure. When fed mouse blood, the seeded cartilage cells multiplied and slowly metamorphosed into the shape of the disappearing scaffold. Vacanti had successfully grown a human ear on the mouse's back. In theory, this meant the method could be applied to humans.[8] Patients could repair their bodies through their own regenerative abilities. Humans could now grow themselves a new pair of ears.

---

Charles is one of four brothers Vacanti who are famous around the world for pioneering the field of tissue engineering.[9] Joseph, Martin, Francis, and Charles all envision a world in which severed spinal cords and diseased organs can be replaced as effortlessly as the transmission of a decrepit Chevy. The boys grew up in Omaha with four other siblings. Their father, Charles Vacanti senior, was a dentist and an academic, and their mother, Joanne, was a mere six hours short of a pre-med degree, which she abandoned to get married. Charles senior, who passed away from a heart attack in 1994, devoted his life to science. Joanne taught her sons that they had been born to help others. Both parents encouraged them to investigate everything in life, even the most outlandish of things. Joseph, the oldest son (known as Jay), dreamed of attending Harvard. But his dad taught at Creighton University, and Jay went there instead, for free. He later made it to Harvard, as did his brothers, who followed the same path.

Jay had wanted to become a doctor since he was four. He eventually specialized in pediatric surgery at the Massachusetts General Hospital, one of Harvard Medical School's teaching facilities. "I have always thought that being a pediatric surgeon was the most gratifying kind of surgery," he says. "You start with the most helpless and vulnerable of humans, diagnose a potentially harmful condition, definitively manage it, oftentimes with surgical intervention, and return the child to his family and to possibly another 80 years of life."[10] Yet early in his career, Jay discovered what would be his biggest obstacle: the frightening shortage of human organs for transplants. Years later—in 1993—Jay and fellow tissue engineering pioneer Robert S. Langer would do a joint study that found that more than US$400 billion is spent annually in the United States on patients with organ failure or tissue loss. This accounts for close to half the country's health care costs. At that time, there were fewer than 3,000 liver donors for the approximately 30,000 patients who die from liver failure. They discovered that 4,000 people die every year while waiting for an organ transplant, despite the eight million procedures performed annually. In fact, 100,000 people die who never even make it to a waiting list.[11]

Back in the early 1980s, Jay already knew that there weren't enough surplus organs in existence. So he decided he would manufacture new ones. He contacted an old research colleague, Langer, a biomedical engineer at the Massachusetts Institute of Technology. Langer was already famous in the field of tissue engineering. At 27, while studying with Jay at Harvard Medical School, Langer filed for his first of hundreds of patents, for a protein-filled plastic he had invented to release medication into the body. He later became famous for overhauling the treatment of brain cancer by creating an innovative time-release method to deliver chemotherapy drugs to the brain without harming the rest of the body. Langer enclosed the drug molecules inside a polymer whose pores controlled the drug release at a slow, steady rate. Approved by the FDA in 1996, it was the first new treatment for this type of cancer in 25 years.[12]

Jay and Langer were both excited by the prospects that tissue engineering held for organ regeneration. They knew of other scientists who had coaxed skin cells to multiply in the laboratory. Skin is an ideal target for tissue engineering; it harbours cells that have regenerative powers, and these cells do not express certain proteins that often cause rejection by the immune system. In 1979 Eugene Bell, an MIT engineer, had successfully grown enough cells to form a blanket of skin. Why not use the same method to grow organs like the liver? The only difference was the specs: skin is essentially two-dimensional, while organs are three-dimensional. The pair spent more than a year trying to figure out how to thicken their engineered tissue, which was as thin as a dime. It was one thing to coax cells to grow horizontally. It was quite another to convince them to transform into a three-dimensional structure with its own circulatory system.[13]

The cells needed a skeleton to guide them. Jay speculated that a scaffold could help them take shape: "I thought degradable plastics would make an ideal scaffold. I knew from my work in cell biology that cells adhere to plastic dishes for in vitro culture, secreting their own scaffolds as they settle to the bottom of the plate," he says. "I also knew you could treat the surfaces of plastics so they would be more likely to cause cell adherence."[14] Not everyone agreed with him, yet Jay soon found that his cells would indeed cling to the plastic surface of the scaffold, where oxygen was available to them. But the inside of the scaffold, where the cells starved of oxygen, remained empty.

Then one summer day in 1986, Jay sat with his four children at the beach in Cape Cod, watching them play in the surf. His gaze became tangled in a web of seaweed. He watched as the slimy, fibrous filaments bobbed along the surface of the water, the wispy green arms probing for nouishment. Nature is brilliant, he thought. A plant's branches stretch out in many directions, searching for nutrients that are then funnelled back to the rest of the plant. Jay's scaffolds needed this interconnectivity. The three-dimensional branches in his scaffold would guide his cells and help them grow. Jay called Langer to see if they could create such a polymer structure. Langer agreed that it seemed possible. And sure enough, it was.[15]

Today, tissue engineering still relies on the principles of the scaffolds developed by Jay Vacanti and Robert Langer. These scaffolds must be biodegradable to allow the host cells to reclaim their job. They must be porous so the cells can weave their way through the insides. They must also be strong enough to maintain the desired shape. Only if the scaffold has all the right chemical, mechanical, and physical properties can the body restart its regenerative clock and renew its own damaged tissue and organs.

---

In 1994, Charles and Jay challenged the FDA with a novel tissue engineering procedure. Twelve-year-old Sean McCormack, from Norwood, Mass., had a condition known as Poland's syndrome; he had been born without chest bones on his left side. There was nothing solid, not even cartilage, protecting McCormack's heart. If he was bare-chested, you could see it beating against his skin. Yet the youngster was one of the best pitchers in local little league baseball, and he refused to let his condition keep him off the mound.

Jay and Charles, along with Dr. Joseph Upton and Dr. Dennis P. Lund, were brought in by McCormack's doctor. The Vacanti brothers scraped some cartilage cells from the boy's breastbone, which jutted out from his chest, and seeded these cells onto one of their polymer scaffolds. "The procedure was so experimental that none of the polymer companies would give us [custom-designed] material for fear of a lawsuit," remembers Jay.[16] So they created their own scaffolds, using polyglycolic acid, which is used for sutures. The round structure mimicked the shape of McCormack's chest wall. Growth factors were added, and the scaffold, which was about the size of a compact disc, was put in a bioreactor. In just a few weeks, the Vacantis found, the scaffold was ready to be implanted.[17] At the Children's Hospital in Boston, the scaffold was placed in the cavity in McCormack's chest, where it integrated quite quickly with his own tissue. A year later, McCormack had a normal-looking chest. "It's pretty cool. It looks like something I was born with," said McCormack four years after the revolutionary surgery.[18] The six-foot-tall teenager had gained back his strength and endurance, and was again able to keep up with his friends. Seven years later, McCormack became a star bicycle motocross racer.[19]

---

Building a human heart in the laboratory is the ultimate ambition of tissue engineering, says University of Toronto professor Michael Sefton. In 1998 the director of U of T's Institute of Biomaterials and Biomedical Engineering proposed just that. After a flurry of e-mails and meetings with Jay Vacanti and other leaders in the field, Sefton spearheaded the Living Implants From Engineering (LIFE) initiative. The purpose of LIFE is to engineer a human heart within 10 to 20 years. When the initiative was first put in motion, there were about a dozen scientists and physicians involved. Now there are at least 60 individuals working in laboratories in Canada, the United States, Europe, and Japan, each with his or her own goal that will bring the initiative closer to fruition.[20]

LIFE is broken down into teams of researchers that focus on different areas of the heart. In Canada, one group of scientists is creating blood vessels to sustain three-dimensional scaffolds. Another group is converting stem cells into cardiac muscle cells, seeding them onto a biodegradable scaffold and generating pieces of heart muscle that could one day be used to patch damaged hearts. Jay and his team are creating valves to connect these patches and incorporate them for more complex applications.

But the $5-billion LIFE project has stalled because of a lack of funding, and is now a loose collective of people working on their own aspects of the initiative and looking for smaller chunks of money to keep their research afloat. Sefton says that trying to raise $5 billion to achieve something as big as engineering a human heart in a decade proved impossible. In the meantime, he says, there are many basic science issues that need to be resolved before the LIFE initiative is completed. For starters, there are the immunology challenges to grapple with. The most utilitarian product would be something that was off-the-shelf, universally applicable for all patients, yet would not be rejected by the body. Taking cells from different individuals is time-consuming, and if patients are elderly—the reality for numerous heart cases—there might not be enough healthy, serviceable cells.

Transplanting a manufactured human heart into a patient will not be the biggest challenge; physicians long ago perfected a method of transplanting donor hearts. But scientists have still not fully mastered the function of cells. When they manipulate cells in experiments, the cells tend to be altered. Cells, then, do not do what comes naturally to them. Rather, they react to situations that force them to respond. It is like letting children grow, says Sefton. "Give them enough direction without trying to control them." Scientists continue to experiment with growth factors and timing, while examining the impact of neighbouring cells. The real difficulty comes far earlier in the process, as they try to manipulate different types of cells to grow at different times, in specific areas of the heart, and to keep oxygen flowing through a three-dimensional scaffold so that all these cells stay alive. Yet what is the heart but a simple pump? There are harder challenges than this in regenerative medicine.

---

John Davies is a real bonehead. Perpetually clad in jeans and comfy pullovers, the tall, burly professor confesses, "I've been a bonehead for some time."[21] Jed, as he is known by everyone, has rosy cheeks, a Welsh lilt, and a puckishness that admits that he likes to play as hard as he works.

In his University of Toronto laboratory there are memos to fellow "boneheads," graduate students who share his passion for all things skeletal. The shelves in Davies' office are crammed with well-loved books on bone. His laptop displays close-ups of the cobwebby, porous material he is obsessed with replicating.

Quiz him on the intricacies of the human skeleton, and Davies is likely to pull out a box of bones left over from his days as a student of dentistry. Rustling around in the dusty box, he may grab a femur and tell you about its hard, smooth outer shell, and the spongy-looking inside that is the marrow. Everyone knows what marrow is; it's the stuff you use to make gravy in the kitchen, he says, tapping the dry, brittle, coral-like substance inside, so wet and sticky and full of goodness when it is alive. Or he may dip into his box of treasures and retrieve an iliac bone, found in the hip, to explain how a bone graft works. Or he may pluck a stray dental implant from the papers fanned across his desk, making you shield your mouth as he describes how an implant is screwed into the gums to replace a tooth's decayed root.

Davies grew up worlds away from Toronto. Wales is a land of song and poetry, Roman castles, and picturesque green hills that slide into the crashing sea. In south Wales, he says, mining, the country's main industry, has created coal-scarred valleys. Even though only two mines now survive, when Davies was growing up most Welsh boys envisioned themselves in soot-covered overalls rather than pristine white lab coats. There were no scientists in the Davies family. Neither parent had gone to university. But it was expected that Davies and his sister would be the first generation to do so. Davies says he was a "waster" at school, and got into university "by the skin of my teeth."

But once he got himself in the door of the Welsh National School of Medicine, Davies says, "I was completely seduced by this environment. I loved it from the first day. I went into the anatomy department and I thought I'd found paradise." Davies aspired to be a doctor, but his grades stifled this dream. Instead, he set his sights on dentistry, which was easier to qualify for. It didn't take him long to excel. "It wasn't work any more. It wasn't the struggle in school," he admits. Davies was offered the chance to earn a degree in anatomy, an opportunity that was usually reserved for medical students, and jumped at it. After he graduated in anatomy, he returned to dentistry with a new-found passion for, and understanding of, the human form.

Davies then made his way to London, where he practised as a dental surgeon for four years, fixing broken jaws, bullet and knife wounds, and wisdom teeth. At night, he couldn't stop thinking of the scientific challenges posed by his patients. While clinical work seemed to keep him busy and content, there was a side of him that missed tackling science in the laboratory. Rather than simply practise dentistry Davies felt he needed to help advance the field. For the first time in his life, the 26-year-old boarded a plane, and headed to Paris, where he worked in a laboratory for three years.

After the research post in Paris had concluded, Davies returned to the London Hospital Medical School's dental anatomy department, where he stayed for seven years. From 1981 to 1988 he taught anatomy and histology (microscopic structure) of bone, and immersed himself in the field of biomaterials. The work was exciting, but the lack of funding was a constant frustration. His salary was frozen, and even though he was the highest grant earner in the department, the funding was far from adequate. So in 1988, when he heard of an assistant professor position at the University of Toronto, Davies decided to pick up his life and move to Canada.

Davies' office is steps away from Molly Shoichet's, in the U of T's Mining Building. When Shoichet is not coaxing nerve cells to crawl across her cellular bridges, she is working with Davies to regenerate old and broken bones. Rather than replace a huge chunk of bone, they prefer to use the body's innate regenerative capabilities to do the work for them. They are encouraging new tissue to grow inside a biodegradable scaffold that temporarily sleeps within the body; the scaffold will disappear when it is no longer needed, avoiding the risk that the body will reject the implant. This new tissue will connect with the surrounding tissue. Since the patient needn't take drugs forever to suppress the natural immune response, this cuts costs, as well as the danger of infection.

Although he spends most of his time in the lab, Davies constantly relates his research back to the clinic. "In this field, where none of this makes any sense unless you get it into patients down the road," he says, "you've got to have some cognizance of what it means to put it in a person." Five years ago, he predicted that they would be in clinical trials within two years. Some of the applications are nearing human studies, but he anticipates inevitable setbacks on the way to the clinic.

---

For decades, scientists have been experimenting with bone's remarkable ability to rebuild itself. In 1965 Dr. Marshall R. Urist revealed that powdered bone implants could propel bone cells to grow and form new areas of tissue. At the University of California, Los Angeles (UCLA), the orthopedic surgeon was experimenting with the collagen matrix that bone clings to, where it naturally grows. Urist took demineralized bits of rabbit bone and inserted them into muscle. New bone formed. He concluded that the matrix housed a regenerative substance that brought bone to life.

In the 1970s, scientist Hari Reddi began a 25-year hunt for this regenerative component. Reddi and his NIH team finally identified bone morphogenetic protein (BMP), which encourages the growth of osteoblasts, cells that have the ability to produce bone. Reddi led the historic experiment that isolated the first of these magical proteins: BMP-7. We now know that there are more than 15 of these proteins. They have been isolated by scientists, and manipulated to make bone grow.[22]

Scientists also know that BMPs are found in tissue other than bone, perhaps aiding in the construction of the kidneys, nervous system, liver, lungs, and heart.[23] Some tissue engineers have seeded BMPs onto scaffolds made of polylactic acid (PLA) or polyglycolic acid (PGA), both used to make sutures. The BMPs must be placed strategically to allow these proteins to contact progenitor cells—cells that are already on their way to becoming specialized types of cells—which then transform into bone-producing osteoblasts. There have been countless studies of BMPs added to different types of matrices to determine which are most effective in the restorative process.

During our lives, bone is continually being replaced and renewed by osteoclast and osteoblast cells. Osteoclasts get rid of unwanted bone, while osteoblasts form new bone. Tissue engineering aims to stimulate this process of osteogenesis (the formation of bone). All we have to do, says Jed Davies, is introduce the right element in the right place within the body, and the work will be performed naturally.

When someone has a broken limb, provided doctors are able to join the two pieces of fractured bone, they simply put the limb in a cast and rely on the body's osteoblast cells to mend the break. If the two bone ends are prevented from wobbling apart, the bone can usually be trusted to repair itself. In a healthy person, tissues on either side of the break will reconnect so seamlessly that within a few years it may be impossible to pinpoint the fracture. But as we age, the balance between osteoclasts and osteoblasts is tipped. Bone doesn't always form as quickly and evenly, and some people become osteoporotic, with a decrease in bone mass and density. If such individuals are in need of a bone graft, they generally have to find a compatible donor, because their own bone no longer stores enough bone-forming osteoblasts and nutrients.

Serious bone defects that cannot be fixed with a cast are at present treated with either an autograft (from the patient's own body) or an allograft (from someone else). For an autograft, the physician cuts a window—usually in the hip—and scoops out some trabecular bone, a spongy type of bone inside the hip bone. This mushy substance is full of marrow, blood vessels, and a bit of fat, depending on the age of the patient. Rather than taking a huge, unwieldy piece of bone to restore the area, the surgeon implants tiny chips of bone that are smothered with cells. The autograft has several drawbacks. As in spinal grafts, the already weakened patient must recover from two surgical wounds—extracting bone to transplant is more agonizing than the pain at the original site of injury, says Davies. Older patients run the risk of donor-site morbidity, he adds, explaining that they may not have enough healthy bone left to transplant. Tissue becomes more fatty as we age, diminishing the amount of usable bone.

If the body cannot provide the raw materials for surgery, an allograft must be used, usually from a cadaver or bone bank. Bone banking dates back to 1950, when Dr. George Hyatt created the first-ever such bank, in Maryland's National Naval Medical Center, because there was not enough bone available to treat the facial wounds of soldiers injured in the Korean War. Hyatt, an orthopedist, helped perfect freeze-drying methods to preserve bone.[24]

While bone from an outside source is treated to minimize the risk of infection, a patient's immune system may still be compromised. In addition, there is often a shortage of serviceable bone, and tissue removed from cadavers does not house as much in the way of growth factors as live samples do. Physicians often rely on artificial materials such as metal plates that can be implanted in the body. Since these synthetic implants do not grow or adjust for growth, some people must have several operations over the years, to adjust the size of their implants. And there are other problems with synthetic implants. They can migrate within the body, causing inflammation. In a hip replacement, if the metal used is too mechanically stable, the patient's surrounding bone may decide that it is no longer of use, and stop working.

Tissue engineering is an alternative method of repairing brittle, splintered bone. Davies and Shoichet are creating scaffolds that convince the body to switch on its restorative cues and grow new bone cells. Achieving ideal cell penetration inside and outside the scaffold requires an ideal environment for growth. A scaffold's material greatly affects whether it will be accepted and incorporated into the body. Bone itself is made up of the polymer collagen, mineralized with calcium phosphate. For years, tissue engineers have attempted to mimic this mineral. They have experimented with cornstarch-based polymers, collagen, coral, and chitosan, which is extracted from the chitin found in crab shells. Coral seems to best mimic bone. The Food and Drug Administration has approved the use of marine coral modified into hydroxyapatite, the mineral component of bone, to repair bone damage caused by "long bone cysts" and "tumour defects."[25] Calcium phosphate has also been used. Many scientists have avoided soft materials that they felt could not bear enough weight. Ceramic, the same stuff our coffee mugs are made of, was introduced in laboratories because of its strength, but this hard, brittle material degrades very slowly, and tissue engineers want their scaffolds to stay in the body only as long as is absolutely necessary.

A scaffold's porosity is also important for good cell growth. Scientists continue to experiment with pore size, shape, interconnectivity, pore-wall thickness, and the rate at which the porous scaffold degrades. There are two types of pores: open and closed. Think of the foam seat-cushion in your car. Each of the cushion's air pockets is completely surrounded by a polymer so that you don't sink through the seat; this is a closed structure. If the cushion's pores were open, they would be interconnected and the air would escape, letting you sink to the frame of the car. An open cell structure like this is ideal for scaffolds, because the cells can communicate with other cells in surrounding pores. Davies and Shoichet's scaffolds are open cell structures with large interconnections between the pores.

While other researchers use salt to create the size and shape of the scaffold's pores, Davies and Shoichet wanted a material that was organic. One of their Ph.D. students devised a method of making polymers with sugar crystals by reducing the temperature and then increasing it during processing, which freezes the polymer and creates the strut-like structure. They called the result Osteofoam, and its patent is secured by their company, BoneTech. But during experimentation, they discovered a problem. They seeded two million cells in a small piece of foam and left them to grow for three weeks, and when Davies and crew returned, he says, they found a nugget of foam. The cells had contracted, and they had crushed the foam, destroying its open pore structure. To get around the problem, they coated the inside of their scaffold with calcium phosphate. This strengthened the structure and prevented the cells from contracting.

---

Making a scaffold is an exacting and arduous process, as Ph.D. candidate Jeffrey Karp will attest.[26] As Karp walks down the exposed brick halls of the Mining Building, the wooden floorboards creak and sigh underfoot. In the laboratory, a pane of glass displays candid photos of colleagues

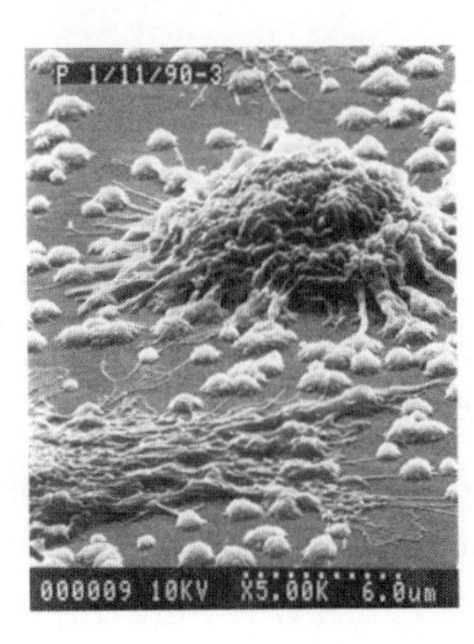


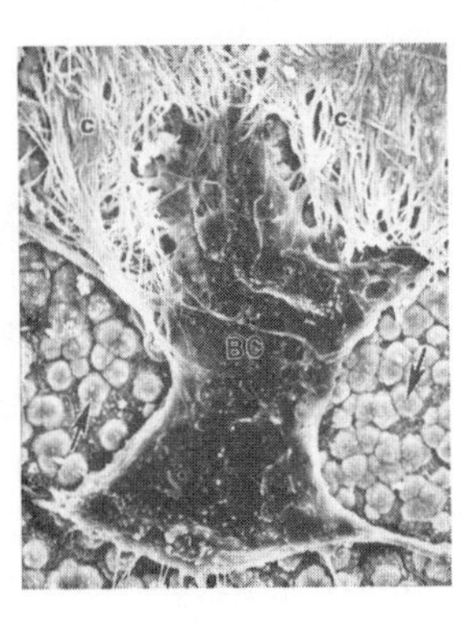


**These images illustrate the biological cascade of early bone formation that occurs within all of us as well as along the surfaces of lab-made scaffolds.**

–Professor John E. Davies

working and celebrating together. In the back of the bustling lab we find a quiet, cool room. Karp takes out a bag of biodegradable plastic beads made of polylactic acid. This material has been approved by the FDA for a variety of applications, including scaffolding. The polylactic acid, a polymer which is carbon-based and thus organic, is the same material used in biodegradable sutures. Karp measures out the polymer and stirs it into a beaker of clear, toxic solvent that is used for drug delivery. The next day the beaker contains a viscous solution, and it is time to add the calcium phosphate, an inorganic white powder resembling chalk dust that strengthens the scaffold.

Karp's next step is to build a mould out of a Teflon-type material similar to the surface of some non-stick fry pans. The material is bendable, and he folds up the sides of the non-stick sheet, creating a small square container about a centimetre high. The outside is aluminum but the inside must be non-stick so the solution can be removed easily. Karp takes 1.18mm granules of sugar—four times bigger than regular table sugar crystals—and passes them through a sieve. Then he takes another sieve that has .85mm holes. Using this method, the pore size he creates for his scaffold is between 1.18 mm and .85 mm. Karp next takes the selected sugar and pours it into the mould, filling the box to the top. The sugar is placed into a humid environment to encourage the surface of the crystals to connect to each other. The joined granules are then left to dry for a day in an oven.

In the next stage, Karp takes the sugar-filled mould and pours in his polymer solution, which flows through the air spaces between the particles of sugar. The mould is then placed in the freezer at –20°C. Within an hour the contents are frozen. He removes the solid block of solution from its mould by releasing nuts on the sides of the box, and places the block in water at room temperature. The polylactic acid doesn't particularly like water, and precipitates out of it. The clear, toxic solvent and the water combine. The result is a block made of polylactic acid and sugar. In time the sugar dissolves, and Karp is left with a polymer scaffold that has pores the size of the sugar crystals. The scaffold is then coated with calcium phosphate. Using cutting tools that he designed specifically for this application, Karp creates cylindrical scaffolds from the block of polymer.

Creating the scaffold is just the beginning of the process of growing bone in the lab. Karp must also create the medium in which the scaffold will swim. In a Petri dish he combines nutrients and vitamins that cells need in order to grow, as well as the steroid dexamethasone, which encourages cells to differentiate. He also adds something to give the medium a long shelf life, and antibiotics to prevent microbes in the lab air from growing in the dish. He takes a rat's femur (thigh bone), cuts off the ends, and flushes out the marrow; this bony mush is dropped into a vial of antibiotics, and is later rinsed with a neutral medium. The marrow is then placed in a Petri dish coated with a solution that encourages cells to adhere to its surface. There are many types of cells within the Petri dish, including stem cells, progenitor cells, muscle cells, and endothelial cells, a component of blood vessels. Growth factors are added to the dish of marrow to encourage undifferentiated cells to turn into osteogenic (bone-making) cells.

The cells are left in the medium for three days. Then Karp rinses them with a solution that washes away those that are not adhering to the dish; experience suggests that they will not interact with growth factors. On the third day he removes the medium from the dish, using a pump, and adds fresh medium. He is looking to separate the committed cells—those cells that are already a particular type of cell—from progenitor cells, which can metamorphose into many different cell types. Washing the cells again, he can purify the target cells, releasing them from the Petri dish. He takes his separated cells and gives them a bath in another test tube filled with a washing solution. Placing these cells in a centrifuge, he causes them to form a pellet and stick to the tube's surface. Once the medium is removed, he is left with all the adherent cells—his desired progenitor and committed cells.

Karp then mashes up his solution and adds some more medium in order to count the suspended cells. Adding a dye, he stains any dead cells, so he knows which ones not to count. (To minimize the variables within a test, he must use a consistent number of cells.) The healthy cells can now be seeded onto Karp's scaffold and left to grow. If he wants to make collagen, Karp will add ascorbic acid (vitamin C). To encourage the collagen matrix to mineralize, he adds a form of calcium phosphate, which mineralizes bone and makes it hard. The cells attach to the scaffold's pores, divide, and proliferate. They grow inside the scaffold and along its outer surface. Mesenchymal cells, which can transform into connective tissues and blood and lymph vessels, are also responsible for producing the matrix; they release neccessary protein. One brick at a time, these cells produce one layer of matrix after another.

Davies and Shoichet's lab-grown bone is so close to the real thing that the trained eyes of many scientists cannot distinguish between micrographs of engineered and natural bone. Yet even with the perfect scaffold, scientists must still devise a method of producing safe, healthy cells—and lots of them. Today, cells are taken from an animal or human and multiplied in the lab. But the safety of using animal cells within people is still being debated. And even if human cells are used, the body may reject them. Some scientists are experimenting with embryonic stem cells, but the mechanisms that control these cells remain a mystery, making them difficult to rely on. Progenitor cells, which are already becoming specific types of cells, are another possibility. Arnold I. Caplan and his team at the Cleveland Clinic have isolated progenitor cells from bone marrow to transform them into bone-forming osteoblasts. Other scien-

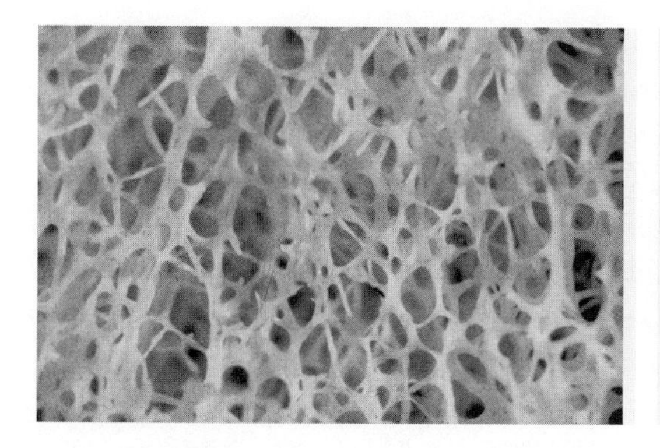

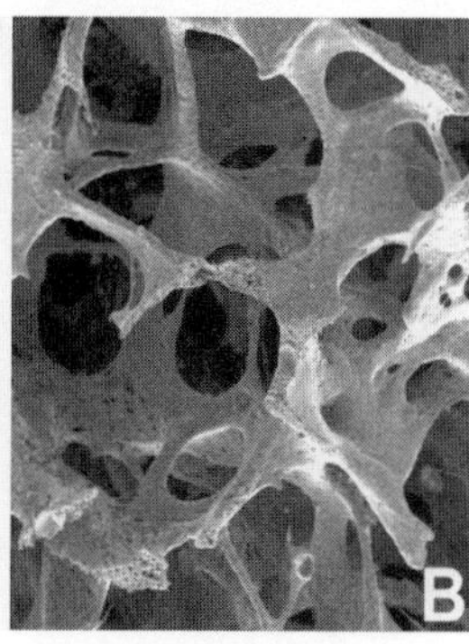


**This scaffold (right), constructed by professor John E. Davies and team, is designed to mimic the architecture of natural bone (left), encouraging bone formation within the body.** —Professor John E. Davies

tists are working on "universal donor" cell lines that would be compatible for all patients and could be mass-produced. The goal is to eliminate or hide the proteins that identify cells as foreign to the host.

As Jed Davies scales up his studies to experiment on bigger, more complex animals, one of the immediate tasks will be to match the rate of scaffold degradation to the speed of bone formation. Scaffolds, he says, will have to be engineered to degrade at different rates in different sites, because bone structure varies throughout the body. For example, marrow from long bones like the femur is more metabolically active than bone in the spinal column.

But even if Davies' scaffolds look, feel, and act like bone, it doesn't necessarily mean that host bodies will accept them. One of the biggest challenges of bone engineering is keeping cells alive during transplantation. In the lab, scientists may be able to multiply cells by the millions and seed them onto a scaffold. But once a patient is cut open to receive the transplant, these cells run the risk of death. Blood may seep into the scaffold, creating a stagnant blood clot. Nutrients may stop flowing to the tissue, which begins to break down. Oxygen may not reach the seeded bone cells, which inevitably suffocate. In order for a cell to survive, it must be within 200 microns of a blood supply. One micron is one-millionth of a metre, so 200 microns is about a hair's breadth.[27]

When the transplanted bone is connected to a blood supply in the host, oxygen-free radicals are produced. These radicals bore holes in the membranes of cells, killing them. At present, surgeons rely on certain chemicals to absorb the radicals. Tissue engineers need to determine more appropriate chemicals to protect their manufactured tissues and organs.[28] While not all the cells must survive, scientists must figure out a way for the patient's body to bring a new blood supply to enough cells to keep the tissue alive and breathing. One alternative is to implant a scaffold that is seeded only with growth factors. Rather than starting the bone-forming process in the lab, these factors could initiate growth within the body.

Regenerative medicine is about improving life. It is about replacing dead tissue and exhausted organs. It is about flipping the biological switch of disease genes to get them to function like normal genes. It is about using the body's own stem cells to repair itself. What, one might wonder, could be wrong with that?

Just ask the Pope, or the president of the United States.

Chapter Six

# Heal Thyself

## When Religion, Ethics, and Science Meet

New life for those whose organs are failing and bodies are deteriorating also means some loss of life. When embryos are mined for their stem cells, these five- to seven-day-old raspberry-like balls die—which some people, including President George W. Bush, believe is morally reprehensible. This despite the fact that these stem cells are removed from embryos donated for research, typically by fertility clinics that no longer have use for the material. While these embryos are headed for the trash—literally—once a woman becomes pregnant and no longer needs them in reserve, controversy remains over the status of the embryo.

There is of course much debate over when life begins, and when a soul—if such a thing exists—comes into being. Is it at the moment of procreation, or when an embryo develops into a fetus (end of the eighth week), or upon a baby's first breath of air? Judaist faith states that an embryo is "like water" for its first 40 days of growth. After this day the fetus is regarded as a human, but "full personhood" does not come until the baby leaves its mother and enters the world as a separate being. Judaism, then, would seem to allow for abortion, another life debate. In the Catholic faith and certain Protestant faiths, the soul is bestowed during conception, which means that embryos and fetuses are both as morally significant as people. As a result, these religions view abortion as premeditated killing, and denounce any destructive scientific research on either embryos or fetuses. The Sunni and Shia schools of Islam espouse the belief that the gaining of personhood is a process that happens over time, that the soul comes into being during the fourth month of pregnancy, when the first fetal movements are witnessed, meaning that embryonic research does not destroy a human soul. Buddhists believe that life starts during conception, with the combination of sexual intercourse, the woman's fertile period, and a Gandhabba, a "being to be born," which is now on the path to life.[1] Such religious perspectives have brought passion and fury to the subject of embryonic and germ stem cell research. Critics cry that researchers are killing future human beings, disregarding the sanctity of life for the progress of science.

Yet others believe that an embryo is in itself no more than a clump of cells. Along the way to humanity, these cells could be ruptured or destroyed by numerous causes. For molecular biologist Lee Silver, "the natural destiny for a human embryo is death." As Silver argues, "The normal reproductive biology of human beings is such that 75 percent of all naturally fertilized eggs will succumb to death naturally before the nine-month period of gestation is completed. It is the odd egg only that develops into a live-born baby."[2]

There are also critics of embryonic stem cell research whose views are not dictated by religious doctrines. Personal ethics can be shaped by cultural and moral codes; after all, the word "ethics" is derived from the Greek *ethos,* meaning the customs of a society. We rely on our ethics to help us apply moral judgements to real-life situations.[3] Our ethics are determined as much by what we believe as by what we don't believe. We separate ourselves morally from our peers by taking a unique and personal stance on issues. Philosopher Annette Baier believes that "Morality is the culturally acquired art of selecting which harms to notice and worry about."[4] Author and veterinarian Michael Fox defines ethics as "that brand of philosophy dealing with values relating to human conduct, with respect to the rightness or wrongness of certain actions and to the goodness and badness of the motives and ends of such actions."[5]

However we define ethics, there is no right or wrong way of shaping our own moral perspective, no one prescribed way of formulating our values. Our lives have been touched by different situations. Something may have happened in our past to sway our moral convictions. Say a teenager becomes pregnant and decides to exercise her right to have an abortion, rather than become a single mother before she has had the opportunity to complete her high school education. A decade later, that same woman may feel compelled to go through with another unwanted pregnancy, now that her life situation has improved. This is an example of situational ethics: situations can alter our ethical stance. Political correctness may also dictate what is and is not acceptable.[6] Until 1918, it was ethically acceptable to most men to deny women the right to vote in Canadian federal elections. And that teenage girl who today chooses to have an abortion would not have been able to do so legally only a few generations earlier.

Fox believes that people are motivated by a love and respect of life; that "we have an instinct as conscious and empathetic beings to act with conscience, and that it is the unified sensibility of feeling and reason that is the basis of ethics."[7] Developing our own personal ethics takes time. From childhood to adulthood, Fox says, we learn how to evaluate impartially our behaviour and that of other people, corporations, and institutions, through several perspectives. Using ethics, we scrutinize values and beliefs, and reflect upon the consequences of our ethical decisions and behaviour.

Then there is the ongoing debate between the "deontologists" and the "utilitarians." If we take a deontological approach to ethics, our actions must be based on their inherent goodness, with no consideration of their consequences. Deontology is based on principles alone. Utilitarianism, on the other hand, prefers the action that will result in the greatest benefit possible for the largest number of people.[8] Deontologists, says renowned ethicist Dr. John Dossetor, believe that respect for human life is absolute. Therefore, even a future good for all of humanity, he says, cannot outweigh the deontological duty to protect individual autonomy. What respect, then, do these two philosophies allow an embryo? A utilitarian might think that sacrificing an embryo is ethically acceptable because the resulting research could result in a cure for cancer, Parkinson's, or Alzheimer's, bettering the lives of millions. A deontologist, on the other hand, might oppose embryonic stem cell research because it involves the death of that embryo.

There are numerous examples within society of utilitarianism challenging deontology, says Dossetor. Think of triage in a hospital, where those with a better chance of being saved are given priority over those who have minimal possibility of survival. Or imagine a young, single mother with glomerulonephritis being given a kidney transplant before an obese, elderly woman with Type 2 diabetes and renal failure: the older woman has less chance of a successful transplant, despite the young mother being farther down the waiting list. The fact that it is acceptable to confine people with some contagious diseases, and vaccinate them against their will, exemplifies the utilitarian view that the individual's right to autonomy can often be sacrificed for the good of protecting the many.

There is a standard Jehovah's Witness case that medical ethicists often describe to explain the ethics challenging a person's right to autonomy. A patient whose religious beliefs clash with Western medical treatments has been admitted to hospital. Even though he refuses care, his physician exercises her authority and treats him. As Georgetown-University philosophy professor T.L. Beauchamp points out, "It is morally prohibited to risk death for a patient whose life threatening condition can be medically managed by suitable medical techniques. . . . It is morally prohibited to disrespect a first party refusal of treatment by a patient, unless the refusal is non-autonomous and presents a significant danger to the patient."[9] While there is great respect for a person's autonomy and beliefs, they do not supersede the health of a patient. Physicians must do their job and fulfill their Hippocratic oath, even at the risk of offending someone's ethical or religious sensibilities.

Different societies collectively adopt different schools of ethics. The Canadian and American cultures, for example, vary in how they approach certain issues. Take gun control. While both Americans and Canadians consider respect for human life a fundamental value, the cultures disagree when it comes to an individual's right to bear arms. Political scientist Ted Schrecker and Margaret A. Somerville, founding director of the McGill Centre for Medicine, Ethics and Law, point out that "despite an epidemic of firearms-related deaths and injuries that would be treated as a national public health emergency if an infectious disease were the cause, the individualism of the society is such that many Americans still assign a higher priority to the constitutionally entrenched right to keep and bear arms."[10]

The reason for this, says sociologist Seymour Lipset, is that "the deeply rooted individualism, anti-government orientation and suspicion of egalitarian ideas distinctive to U.S. political culture are legacies of the ideas that guided the American Revolution; in contrast, Canada's political origins were distinctly non-revolutionary."[11] Americans also tend to be viewed as more capitalist-minded. Some American scientists believe in the "ethics of the marketplace": they feel that consumers are ethical individuals who will not purchase something that is unethical, bankrupting those who produce anything immoral. This theory seems hard to accept, considering that the worldwide sale of kidneys and other organs continues on the black market, despite the fact that it is morally wrong. Could it be said that those who believe in the ethics of the marketplace also believe it makes economic sense to enhance and manipulate embryos to improve children's intelligence, memory, and musical ability, because it will cut the cost of private school tuition, tutors, and music lessons?[12]

Françoise Baylis, professor of medicine and philosophy at Dalhousie University, is concerned about the individualistic view of many people who think their rights outweigh the rights of society. "I find too much of contemporary bioethics just pays attention to me and my rights, so that if I can say I have a right to do this and I'm not hurting you, just leave me alone and I'll pay for it." Baylis says that personal decisions inevitably have a societal impact beyond the realm of the individual. Even if the intention is to use something for good, it can also be used for negative and dangerous purposes. In her lectures, she says, "I talk to people about a hammer. I say, If I use a hammer as a gavel because I'm a judge, you would say that's a good use of a hammer. If I use a hammer to build a house for the poor, you would say that's a good use of a hammer. If I take a hammer and bash someone's brains in, you will tell me it's a bad use of a hammer, but you're not going to tell me it was not a hammer." Potential misuse, says Baylis, is not grounds for banning a technology. "You should never say, Oh well, some lunatic out there is going to do this terrible thing, therefore we can't [use that technology]," she explains. "I think that that's not a very strong argument."[13]

Take somatic cell nuclear transfer (SCNT). This controversial and experimental procedure, also known as "therapeutic cloning," creates embryos for research. Somatic cells are non-reproductive cells—not sperm or eggs. In SCNT, the nucleus of an unfertilized egg cell is extracted. The nucleus of a somatic cell—say a skin, heart, or brain cell—is also removed, and inserted in the enucleated egg cell. A totipotent cell results, which scientists can use to create a blastocyst, a very early stage of embryo. In the laboratory, an electric pulse triggers embryonic development. Stem cells are extracted from the blastocyst, isolated, and studied. Through SCNT, millions of stem cells can be harvested without destroying embryos that exist outside research. For years, cloning has inspired science fiction writers and offered surreal movie plots. In 1975 and again in 2004, *The Stepford Wives* invited audiences to a picture-perfect American suburb where wives had been replaced by subservient clones. In *The Sixth Day,* Arnold Schwarzenegger plays a helicopter pilot who returns home after a near-death experience to come face to face with his own clone. Today's headlines make it harder and harder to separate fact from fiction. Technology now allows for cells, genes, even living organisms, to be copied.

Many people fear SCNT because it uses the same technology as reproductive cloning, which results in the replication of a living organism. The manufactured embryo could be implanted into the uterus of an animal or human to create its own clone. Both good and bad scenarios come with such power. On the positive side, parents could clone a deceased child, or a grieving widow could re-create her dead spouse. A patient could create his or her own clone and harvest compatible organs or tissue for desperately needed transplants. Parents carrying a genetic disease could have a child free of the faulty genes.[14] On the negative side, cloning could result in the commodification of people, with individuals viewed as products to buy and sell. Couples could decide they would rather not run the risk of having children born with disabilities, and opt to clone themselves a family. Rather than using the technology to prevent passing on a disease, the clone's genes could be enhanced to produce the "perfect" child, erasing any unwanted traits. In Françoise Baylis' words, it is about "getting it right" in order to sidestep the genetic misfortunes of a previous generation. As Michael Fox explains, "We cannot put our trust in science, for a very simple reason. . . . What we do with knowledge that science creates is not the business of science. Science has nothing to do with good or evil, with the satisfaction of human desires. . . . And so we have to decide for science what is worth doing before we use science to do it."[15]

---

While the UN aims to draft an international treaty to ban reproductive cloning, the lack of regulations has permitted many scientists to continue their experimentation. In 1997 Ian Wilmut and his team at the Roslin Institute in Edinburgh, Scotland, removed the nucleus from a cell in a six-year-old sheep. It was transplanted into an unfertilized egg whose own nucleus had been removed. In a laboratory, the egg was helped along in its development. Two hundred and twenty-seven embryos grew, but only one was brought to full term. The result: Dolly the sheep, the first mammalian clone.

In 1999 a group of South Korean scientists made an unconfirmed report of human cloning from somatic cells. In 2001 Advanced Cell Technology announced it had cloned human embryos.[16] Then, in 2004, the South Korean researchers announced that they had created 30 cloned human embryos using SCNT, grown them to the blastocyst stage, and successfully harvested stem cells for research. If the technique is perfected, therapeutic cloning could offer an endless supply of stem cells to grow replacement tissue and organs.

But not all scientists around the world are permitted to experiment with this technology. In Canada, the Assisted Human Reproduction Act,[17] which came into force in 2004 after a decade of false starts, prohibits human cloning. Subsection 5(1) of the act provides that:

5(1) No person shall knowingly

(a) create a human clone by using any technique, or transplant a human clone into a human being or into any non-human life form or artificial device;

(b) create an in vitro embryo for any purpose other than creating a human being or improving or providing instruction in assisted reproduction procedures;

(c) for the purpose of creating a human being, create an embryo from a cell or part of a cell taken from an embryo or foetus or transplant an embryo so created into a human being;

(d) maintain an embryo outside the body of a female person after the fourteenth day of its development following fertilization or creation, excluding any time during which its development has been suspended;

(e) for the purpose of creating a human being, perform any procedure or provide, prescribe or administer any thing that would ensure or increase the probability that an embryo will be of a particular sex, or that would identify the sex of an in vitro embryo, except to prevent, diagnose or treat a sex-linked disorder or disease;

(f) alter the genome of a cell of a human being or in vitro embryo such that the alteration is capable of being transmitted to descendants;

(g) transplant a sperm, ovum, embryo or foetus of a non-human life form into a human being;

(h) for the purpose of creating a human being, make use of any human reproductive material or an in vitro embryo that is or was transplanted into a non-human life form;

(i) create a chimera [single organism made of two genetically different tissue types], or transplant a chimera into either a human being or a non-human life form; or

(j) create a hybrid [the offspring of two different species] for the purpose of reproduction, or transplant a hybrid into either a human being or a non-human life form.

Section 60 of the same act provides that those who contravene these provisions, which also apply to therapeutic cloning for research purposes, could face up to 10 years in prison and fines of up to $500,000.

In the United States, the House of Representatives passed a bill in July 2001 that makes it a federal crime to clone embryos for therapeutic or reproductive purposes. The Human Cloning Prohibition Act of 2003 was presented to the House on February 27, 2003. The House voted 241 to 155 in favour of the act, which prohibits reproductive and therapeutic cloning. Violators face up to a decade in prison. President George W. Bush has decreed that no federal funds are to be used to create embryos for research purposes or to clone embryos for any purpose. These rules do not apply, however, to the private sector.

Michael Sandel sits on the President's Council on Bioethics. He asks why the creation and sacrifice of excess embryos for fertility purposes is morally acceptable, if the development of embryos for stem cell research is viewed as morally wrong. Both have noble intentions. One is to give life; the other is to save lives through regenerative medicine. According to Sandel, "curing diseases like Parkinson's, Alzheimer's, and diabetes is at least as important as enabling infertile couples to have genetically related children." He goes on to state, "Opponents of research cloning cannot . . . endorse the creation and use of excess embryos from fertility clinics and at the same time complain that creating embryos for regenerative medicine is exploitative. If cloning for stem cell research violates the respect the embryo is due, then . . . so do in vitro fertilization [IVF] procedures that create and discard excess embryos. . . . [T]he moral arguments for research cloning and for research on leftover embryos stand or fall together."[18]

Other members of the President's Council on Bioethics think otherwise. A section in its report entitled *Human Cloning and Human Dignity: An Ethical Inquiry* states, "In the eyes of those who create IVF embryos to produce a child, every embryo, at the moment of its creation, is a potential child. Even though more eggs are fertilized than will be transferred to a woman [in a given cycle], each embryo is brought into being as an end in itself, not simply as a means to other ends. Precisely because one cannot tell which IVF embryo is going to reach the blastocyst stage, implant itself in the uterine wall, and develop into a child, the embryo 'wastage' in IVF is more analogous to the embryo wastage in natural sexual intercourse practiced by a couple trying to get pregnant than it is to the creation and use of embryos that requires (without exception) their destruction."[19]

Not all states are in agreement with President Bush's decree or his bioethics council. California passed legislation that allows for creation of and experimentation on cloned embryos. A New Jersey law permits somatic cell nuclear transfer, but a related New Jersey law prohibits cloning, which is defined as "replication of a human individual by cultivating a cell with genetic material through the egg, embryo, fetal, and newborn stages into a new human individual."[20] This definition appears to permit therapeutic cloning for research purposes while banning reproductive cloning, even though both use the same technology.

In Austria the Reproductive Medicine Law of 1992 permits the creation of embryos, but only for reproductive purposes. In France the creation of embryos for scientific objectives is prohibited. The United Kingdom's Human Fertilisation and Embryology Act 1990[21] allows embryos to be created and grown for research purposes for up to 14 days. The reasoning behind this is that for the first two weeks the embryo is merely a cluster of cells. On day 15 the body of the embryo is created, enabling its head and tail to be identified. Also, after two weeks, implantation of the embryo into the womb is finished.[22] Yet the U.K. legislation prohibits one type of cloning: "nuclear substitution of any cell whilst it forms part of an embryo. . . . The cell nuclear transplant technique involves nuclear substitution into an egg not an embryo."[23]

In Singapore the Bioethics Advisory Committee's 2002 *Ethical, Legal and Social Issues in Human Stem Cell Research, Reproduction and Therapeutic Cloning* suggests that, "The creation of human embryos specifically for research can only be justified where (1) there is a strong scientific merit in, and potential medical benefit from such research; (2) no acceptable alternative exists; and (3) on a highly selective, case-by-case basis, with specific approval from the proposed authority."[24]

The Bioethics Advisory Committee of the Israel Academy of Sciences and Humanities report entitled *The Use of Embryonic Stem Cells for Therapeutic Research* permits therapeutic cloning, stating that "the Israeli Law on Genetic Interventions in Humans, while prohibiting the creation of a 'complete human being' by reproductive cloning, does not rule out producing cloned embryos that will not be implanted." Japan's 2001 Law Concerning Regulation Relating to Cloning and Other Similar Techniques prohibits cloned embryos from being implanted into a uterus, while permitting these embryos to be used for research purposes.[25]

Cloning for reproductive and therapeutic purposes raises many ethical issues. Religious groups contest that cloning humans or improving the genome we were born with has scientists playing God, disrupting the natural order of things by manipulating humanity, and destroying God's master plan. The U.S. National Bioethics Advisory Commission's (NBAC) report entitled *Cloning Human Beings* states, "Human beings should not probe the fundamental secrets or mysteries of life, which belong to God. Human beings lack the moral authority to make certain decisions about the beginning or ending of life. Such decisions are reserved to divine sovereignty. Human beings are fallible and also tend to evaluate according to their narrow, partial, and frequently self-interested perspectives. Human beings do not have the knowledge, especially the knowledge of outcomes of actions attributed to divine omniscience."[26] Yet, as Françoise Baylis points out, it could be argued that God expects individuals to rely on their reason, imagination, and freedom to better their quality of life.

Testing religious faith is the belief in the powers of science. In the late 1500s and early 1600s, Francis Bacon, known as the father of biotechnology, believed humanity should use its power over nature to improve quality of life. Bacon believed that knowledge was power, and that the acquisition of knowledge should be the ultimate goal in life. This made it morally acceptable to change or harm nature in the process. In one of Bacon's last publications, he muses on the power of science to improve life. "We create worlds. We prescribe laws to nature and lord it over her. We want to have all things as suits our fatuity, not as fits the Divine

Wisdom, not as they are found in nature. We impose the seal of our image on the creatures and works of God, we do not diligently seek to discover the seal of God on things."[27]

---

It is difficult for governments to regulate science, which is constantly changing and challenging society's ethics. Government bodies that create and enforce laws must ensure that the laws are ethically acceptable to the general public. While it is impossible to please everyone, governments and lawmakers must work with the values and morals of society, and indeed humanity as a whole, in mind. The ethical concerns of people must be identified, and conflicting ethical perspectives must be addressed, in order to legitimize national policies regarding stem cell research. It is imperative that policy-makers separate ethical perspectives from financial or political interests. They must distinguish whether people are voicing their personal, entrenched values, or a point of view that could improve their position within society.[28]

Political scientist Ronald Inglehart writes about a materialist versus post-materialist view of life. Materialists base life decisions on financial security and view economic growth as a political priority. Post-materialists prefer to focus life choices on personal fulfillment, and believe that politics should vehemently protect freedom of speech and human rights. Both camps of ethics seem to have gained favour at different times in history. Inglehart has discovered that in many countries, over time, there is an increase in the number of post-materialists. This, he says, can be attributed to a change in "material circumstances" and the ways in which these circumstances influence "pre-adult political socialization." Most people who grew up in the post-war industrialized world enjoyed a life in which material needs were taken care of and their relationship with money was different from that of their parents who just scraped by during the Great Depression.[29]

Pivotal to a democratic society is understanding how an ethically sensitive policy is reached, rather than simply identifying the outcome. There must be transparency to the process that explains the basis for a decision.

Due process is also necessary to ensure that policy-makers are free of conflicts of interest, and to allow stakeholders an opportunity to state their position. Ted Schrecker and Margaret Somerville advise that governments ought to identify the core values and standards of ethical acceptability of their people. "The emergence of human rights as a 'modern tool of revolution,'" they write, "a category and a set of concerns that can be (and indeed must be) applied across national boundaries while respecting differences in national capacities and cultures, suggests that this can be done effectively. Human rights provide a basis for insisting that (for instance) race-based or gender-based exclusions from opportunities to participate fully in the life of a society were wrong a century ago just as they are today, and for the same reasons."[30]

Deciding not to make a decision is as ethically problematic as enforcing a morally ambiguous policy. Governments are accountable to the public. By remaining undecided and letting an issue hang, governments inevitably take a stance through their very inaction. Take the Canadian government's decade-long debate over stem cell research. In 1993, after four years of studying Canadian activities in the area of assisted human reproduction, the Royal Commission on New Reproductive Technologies released a two-volume report recommending that the federal government ban human cloning and establish an oversight committee to regulate new reproductive technologies. The report, which had 15 volumes of supporting material and discussion, emphasized that this field should be guided by an "ethic of care," and also recommended a national regulatory body at arm's length from the government, comprising at least 50 percent women.[31] Two years later, the Minister of Health introduced a voluntary moratorium on certain ethically sensitive procedures identified by the commission. In 1996 an advisory committee was established to oversee these procedures.

That same year, Bill C-47, the proposed Act Respecting Human Reproductive Technologies and Commercial Transactions Relating to Human Reproduction, was introduced and approved by the House of Commons Standing Committee on Health. But the bill died with the breakup of Parliament in the spring of 1997. In 2001, the Minister of

Health presented the draft legislation to the House of Commons Standing Committee on Health, asking for a review. In 2002 the Minister introduced a second bill, Bill C-13, the proposed Act Respecting Assisted Human Reproduction, with the intention of regulating assisted human reproduction and embryonic research. It wasn't until October 28, 2003, that the legislation passed its third reading in the House of Commons, by a vote of 149 to 109.

In the spring of 2004, in order to move the third reading of the bill forward, more than 50 witnesses appeared before the Standing Senate Committee on Social Affairs, Science and Technology to voice their apprehensions and applause for the bill. Those supporting the bill include Dr. Ron Worton, scientific director of the Stem Cell Network, who has said that the demise of the bill would give private corporations the freedom to research human stem cells. There need to be checks and balances, says Worton. Without proper legislation, researchers and scientists can do as they please. And what if they choose to clone a human? If a bill is not in place banning the activity, there is no one to stop them.

Members of the Jewish and Muslim faiths expressed their concerns about the proposed legislation, but concluded that it was better to have some form of regulation in place rather than none at all. Monsignor Prendergast, the Archbishop of Halifax, recapped Catholic opposition to research on human embryos and assisted reproduction, but could also separate the constructive aspects of Bill C-47. By banning cloning, the proposed law would protect human embryos, despite scientists' right to experiment on leftover embryos from fertility clinics.[32] During the moving forward of the third reading, Senator Yves Morin said, "What struck each of us most was the degree of consensus on such controversial legislation. More than two-thirds of all witnesses recommended passage of the bill without amendment. Despite the reservations many of them had, they believed that legislation in this area is long overdue and should not be held up any further."[33]

Supporters of C-47, like Angela Campbell, assistant professor at McGill University Law School, agree that using criminal sanctions to enforce the legislation is appropriate. There needn't be unanimous public approval on the wrongfulness of an act for it to be banned by Parliament, she says, adding that there are many activities prohibited that not all Canadians feel are morally wrong. Campbell cites two values entrenched in the Canadian Charter of Rights and Freedoms: the inviolability and sanctity of human life, and human dignity. These values, she believes, could be threatened by reproductive technologies. "For example," she says, "the creation of human embryos for use in research devalues this form of human life by allowing the embryo to be objectified and pragmatically used as a means to a separate goal." Criminal law in Canada must be used sparingly, concedes Campbell, for actions that are deemed wrong and harmful. Does this apply to embryonic research and reproductive technologies? "Because activities such as human cloning, transgenic science [mixing genes from different species], and buying, selling and brokering in human embryos potentially thwart the values of human life and dignity, it is my view that they are wrongful and harmful," she writes. "The harm that they threaten to engender is serious enough—both in nature and degree—to warrant the imposition of the criminal law in their regulation."[34]

Not everyone is in agreement with Campbell. Timothy Caulfield, professor at the University of Alberta's Faculty of Law and research director of the university's Health Law Institute, has been very vocal in the stem cell debate. "I think using criminal law in this context is ridiculous," he says. "Having those kinds of punishment associated with something that there is so much moral ambiguity around just intensifies the absurdity, I think." Who would be sent to jail? he asks. The whole laboratory?[35] Criminal bans, says Caulfield, should be used only as a last resort. In 1982, the Canadian government stated that criminal sanctions are reserved for "conduct which is culpable, seriously harmful and generally conceived of as deserving of punishment." Once an act has been given criminal penalties involving a loss of liberty, Caulfield says, it is extremely difficult to lessen those penalties.

"What I think a prohibition does is, it implies closure of the debate. This is law, and it's done," he says. "And you're going to get a very polarized debate just as you've seen around medicalized marijuana, abortion, all these areas where we've tried to use the heavy hand of criminal law . . . the debate happens at the margins, at the extremes. I think what happens is, you lose that middle ground, the more nuanced debate." The problem with such an all-or-nothing attitude towards these technologies is that it treats science as fixed, when in reality it is mutable and evolving. Caulfield feels legislation should reflect such fluidity. Once a law has been enacted, he says, history has shown that it is near impossible to amend it. "You get these laws that are cast in a certain time's view of science, and science moves forward so quickly that [the law] can quickly become irrelevant."

Caulfield also disagrees with the ban on somatic cell nuclear transfer. People say therapeutic cloning is unsafe and commodifies the embryo, he says, but "We don't ban other techniques that facilitate the commodification of human tissue." Just look at kidney and other organ transplants. This slippery-slope argument sees reproductive cloning as a natural extension of therapeutic cloning. Proponents of the ban say the public wants to enforce prohibitions against SCNT. Yet Caulfield says there is no evidence to prove such consensus. On the contrary, he says, the public seems to support the technology. He writes, "In fact, I am unaware of any data that show a majority of the Canadian, British or American public to be against therapeutic cloning. Recently, the U.S. President's Council on Bioethics explicitly noted this lack of consensus stating that, therefore, a ban on all forms of human cloning was not justified and that a moratorium should be imposed to give time 'to seek moral consensus'—a surprising result given that a more conservative recommendation was anticipated."[36] The real reason for the ban on SCNT, argues Caulfield, is to revive the pro-life debate. For a very vocal minority, this legislation offers another opportunity to engage the Parliament of Canada in the politics of abortion. He points out that there is no consensus among religions on the moral status of the embryo, and strongly believes that Canadians shouldn't be held hostage to a single point of view. Despite opposition from many lawyers, scientists, and ethicists, Bill C-6 (formerly Bill C-13 and Bill C-47), banning therapeutic cloning, was finally given royal assent, and became law on March 29, 2004. Senator Morin did, however, concede that SCNT is a promising technology that merits further study when the legislation comes up for review.[37]

The good news about the Assisted Human Reproduction Act, says Caulfield, is that it creates the Assisted Human Reproduction Agency of Canada (AHRAC), a regulatory body responsible for monitoring embryonic research as well as implementing quality control and reporting structures for in vitro fertilization clinics. AHRAC will report to Parliament via the Minister of Health. Its mandate includes licensing assisted human reproduction procedures, inspection of IVF clinics and laboratories, advising the Minister of Health on such reproductive technologies, and monitoring national and international developments in the field. Scientists who have been granted licences to experiment on embryos must demonstrate to AHRAC that it is necessary to use an in vitro embryo for the purposes of their research, and that the embryo was originally intended for reproductive purposes—the embryo cannot be created specifically for research purposes; informed written consent must be obtained from embryo donors, and no money can be exchanged for these embryos.

The banning of SCNT by the Canadian act is not just an ethical concern; it could one day prevent researchers from doing their jobs. There is a shortage of surplus human embryos that scientists can use to advance their research. A study led by Dalhousie University's Françoise Baylis revealed that, as of August 2003, 15,615 embryos were being stored in fertility clinics across Canada, but only 299—about 2 percent—were slotted for research. The donated embryos would offer as few as seven and no more than 36 stem cell lines, or populations, for experimentation.

Baylis and her colleagues identified 26 IVF clinics, but only 13 clinics completed the questionnaire. In the United States almost 400,000 cryopreserved (frozen) human embryos have been discovered, with only 11,283 slotted for research. Another research team, led by D.I. Hoffman, found that if these embryos available for research could be used, 275 stem cell lines would probably result.[38]

American scientists do not have such freedom. As of August 9, 2001, President Bush prohibited the use of federal funding for research carried out on human embryos, the creation of human embryos for experimentation, or the cloning of such embryos for any purpose whatsoever. Prior to this date, scientists had already sacrificed many embryos, creating about 60 stem cell lines. Because, as Bush put it, "the life and death decision had already been made,"[39] government-funded researchers are permitted to experiment on these stem cell populations—if they were derived from excess embryos created for reproductive means, informed consent was given by donors, and those donors were not financially remunerated for their embryos.

The American debate over embryos began shortly after the Supreme Court's landmark 1973 decision in the case of *Roe v Wade*,[40] which legalized abortion across the United States. At the time, concern was already growing over the potential use of aborted fetuses and embryos, so the American Department of Health, Education and Welfare (DHEW) issued a moratorium on funding for research using human fetuses and embryos. In 1975 the National Commission for the Protection of Human Subjects of Biomedical and Behavioral Research created guidelines for such reproductive research, and the moratorium was lifted. In 1978, in Britain, the first child conceived by IVF was born.

The DHEW continued to refuse funding for human embryonic research. Through the 1980s, when the Ethics Advisory Board disbanded, a de facto ban on funding for such research remained in place. In 1993, changes in Congress allowed for the possibility of NIH funding for human embryonic research using donated embryos from IVF clinics. An NIH Human Embryo Research Panel explored the concerns regarding such reproductive studies, and advised that certain types of embryonic research should be eligible for government subsidies. Then President Bill Clinton refused to allow money to go to studies on embryos created for research purposes, but agreed with the NIH panel that funding could support studies using discarded embryos from fertility centres. Congress, however, did not agree.

In 1995, Congress amended the Departments of Labor, Health and

Human Services, and Education, and Related Agencies Appropriations Act, which banned federal funding for the creation of human embryos, or embryonic research that destroyed or harmed human embryos. It was called the Dickey Amendment, after Representative of Arkansas Jay Dickey. The amendment does not affect research that is privately funded; it only prohibits government-funded projects. "Thus, it addresses itself not to what may or may not be lawfully done, but only to what may or may not be supported by taxpayer dollars," states *Monitoring Stem Cell Research,* a report of the President's Council on Bioethics. "At the federal level, research that involves the destruction of embryos is neither prohibited nor supported and encouraged."[41] The Dickey Amendment has been re-enacted annually since its creation.

In 1999, after close scrutiny of the Dickey Amendment, the U.S. Department of Health deduced that its specific wording could allow for certain human embryonic research to be funded. What if these embryos were first destroyed using private funds? The public purse would then be opening for research involving already destroyed embryos. The government could thus advance science while continuing to uphold its abhorrence for the destruction of life. Others said this clashed with the "spirit" and the "principle" of the law, and contested this loophole; by offering funding to study destroyed embryos, they argued, the government would be promoting their destruction.

But the Clinton administration had decided that supporting such advancements in science was a noble pursuit. In August 2000, President Clinton noted, "I think we cannot walk away from the potential to save lives and improve lives, to help people literally to get up and walk, to do all kinds of things we could never have imagined, as long as we meet rigorous ethical standards."[42] Although the administration compiled guidelines to regulate this process, Clinton's term finished before they were ever put into practice.

When President George W. Bush entered the White House, he examined the amendment, searching for a way to uphold its spirit while funding Americans in the worldwide advance of science. He decided to allow research on those stem cell lines already destroyed before August

9, 2001, while forbidding the destruction of any living embryos from that day forward. During the announcement of this amended policy, Bush remarked, "There is at least one bright line: We do not end some lives for the medical benefit of others. For me, this is a matter of conviction: a belief that life, including early life, is biologically human, genetically distinct and valuable."[43] He announced that there were more than 60 "genetically diverse" stem cell populations already in existence for scientists to work with. This sounded like promising news. But by 2003, it had become clear that these cell lines were inadequate, and far fewer than expected. The 60 lines include some that are not available for research, and others held by private corporations. The 12 lines that are available do not offer the genetic diversity needed for research.

Furthermore, these dozen cell populations were all grown using mouse cells and could be carrying mouse DNA, or mouse viruses that human immune systems might be unable to fight.[44] John Gearhart, Ruth Faden, and a team of scientists conducted a study to examine the risks associated with stem cell transplantation—namely, cross-species transfer of infectious disease. While it is common for infections to be transmitted from a donor to a recipient, the mixture of mouse and human cells poses a greater risk, one that scientists cannot yet fully comprehend.[45] The team pointed out that it is impossible to predict how these cells will behave once they are implanted into humans. They could transfer genetic disorders to recipients. And the cells could "misdifferentiate" into inappropriate cell types, or they could miss their target altogether and wreak havoc on other organs or tissues of the body, causing dangerous side effects. In conclusion, the scientists state, "We believe that it is unethical to expose human subjects to the stem cell lines that are currently approved for use in federally funded research."[46] As Faden points out, "Conducting a federally funded clinical trial of human ES [embryonic stem] cells under current federal policy would require using cell lines that none of us feel should be used in people, since it is now feasible to create safer lines."[47]

American corporations have complete freedom to destroy and experiment on human embryos, because they are privately funded. How can stem cell research be morally reprehensible for federally funded scientists, yet perfectly acceptable for private laboratories? If an action or pursuit is prohibited by law because it is considered wrong or unjust by elected officials and the society that elects them, then the unjustness of that action or pursuit should be absolute, regardless of who pays the bills. A report of the President's Council on Bioethics offers this defence: "The decision to fund an activity is more than an offer of resources. It is also a declaration of official national support and endorsement, a positive assertion that the activity in question is deemed by the nation as a whole, through its government, to be good and worthy. When something is done with public funding, it is done, so to speak, in the name of the country, with its blessing and encouragement."[48]

Critics of President Bush's non-position on stem cells are not impressed. "That's the deep paradox of the United States policy in this area," says Tim Caulfield, "because they have in some respects a very restrictive public environment. But on the other hand, they have an incredibly open private environment." John Gearhart, the scientist who first isolated germ (reproductive) stem cells, and James Thomson, who initially isolated embryonic stem cells, were funded by the private company Geron Corp., which owns the exclusive rights to profit from the scientists' discoveries. To avoid politics, nothing in Gearhart's lab, not so much as a light bulb, was bought with federal funds, even though his research was done on aborted fetuses, which federal law permits. Thomson's work involves destroying embryos, which is prohibited by the Bush administration, so he worked in another laboratory, on the other side of the University of Wisconsin-Madison campus, physically separating this controversial project from his NIH-funded research.

Giving the go-ahead to privately funded scientists while clamping down on those funded by the government removes intellectual property from the public domain and hands it over to corporations that then patent the knowledge and profit from its use. Caulfield says President Bush's stance on stem cell research is politically motivated: "The idea is, Bush wants to be able to say to his constituency that no public dollars are going to the creation and destruction of human embryos." According to a report by the President's Council on Bioethics, several hundred researchers at about 10 U.S. companies are experimenting with embryonic stem cells, and spending more than US$70 million. This is much more than double what the NIH has spent in the field.[49]

Frustrated by policy restrictions, scientists are taking matters into their own hands. Harvard University, home to one of the chief centres for regenerative medicine, refuses to be pushed to the sidelines to watch other scientists around the world explore this new frontier of medicine. It has launched the Harvard Stem Cell Institute, which funds all types of stem cell research, including controversial embryonic studies. About 100 scientists are signed on to study these cells, to find cures for diseases including Parkinson's and heart disease. "Every success will change the argument," says Dr. Leonard Zon, president of the International Society for Stem Cell Research. "The American people will not stand for scientists not being able to work on their diseases."[50]

Private donors helped launch the institute, as did organizations concerned with specific diseases, such as the Juvenile Diabetes Research Foundation. There have been many donations from individuals, such as the US$5-million contribution from Howard Heffron, a graduate of Harvard Law School's class of 1951.[51] "Harvard has the resources, Harvard has the breadth, and, frankly, Harvard has the responsibility to be taking up the slack that the government is leaving," says Dr. George Q. Daley, who participated in the launch of the institute.[52] The Harvard institute will offer researchers in competing laboratories access to these stem cell populations. In March 2004, Doug Melton, co-director of the institute, announced that 17 new stem cell lines could be used for free by scientists.[53]

An anonymous US$12-million donation is launching Stanford University's own Institute for Cancer/Stem Cell Biology and Medicine, which will be led by Dr. Irving Weissman, who was the first to isolate hematopoietic stem cells in mice (see Chapter Four). The Stanford institute will create new stem cell lines for research, as well as experiment with therapeutic cloning, putting Weissman once again in the line of fire. Yet creating the new stem cell lines makes the controversy worth it. In fact, the scientist says it would be unethical not to proceed, knowing that such regenerative technology could loosen the grasp of genetic diseases on humanity.

Dr. Thomas Okarma, CEO of Geron, entered the field of stem cells in 1995, three years before Thomson and Gearhart's historic discoveries. He is putting his money on stem cells and their billion-dollar regenerative powers. Okarma has heard all of President Bush's arguments denouncing the stem cell field. Even though Okarma is unaffected by the President's regulations, he chose to testify in front of the President's Council on Bioethics. What he heard appalled him: "The metaphor that these people use to describe embryonic stem cells [is] children who are imprisoned obviously against their will, whose imprisoners kill them for their organs for transplantation."[54]

This despite the fact that embryonic stem cells are taken from days-old blastocysts typically donated by fertility clinics when they are no longer of use. Not to mention the fact that sacrificing a few stem cells could relieve countless people of debilitating and fatal diseases. The debate over human embryonic research, Okarma believes, will "continue to smoulder until we demonstrate an absolute breakthrough in biological medicine. And that's what we think stem cells are going to do in people." The progress of science stops for no one, not even the President of the United States. Stem cell research will one day lead to drastically different medicines custom-designed to our own particular bodies and diseases. "Today's cells," says Okarma, "are tomorrow's pills."

## Chapter Seven

# Golden Cells and Flying Pigs

## The Business of Biotechnology

Science is magic. It uses what is within all of us—our common genes and proteins, the cells that make our hearts beat and our brains compute—to unlock the codes of the diseases that plague our bodies. It understands that we hold the answers to our life's ills. Our own cells can uncover that which makes each of us mortal. Some of us house cancer cells that want to live forever, killing everything in their path. Some have faulty genes hibernating within us, producing renegade proteins that wreak havoc on our respiratory or circulatory systems. Some of us were born with a glitch in a chromosome that is equivalent to a death sentence, though it may not take effect for decades, while most of us have a majority of cells that function normally all of our years, living, aging, and inevitably dying, slowly counting down our days.

Science is a mystery. We now understand that stem cells can become whatever their destiny decrees, be it skin, muscle, or bone. But why do they take one path rather than another, and why at a given moment? Scientists grasp the healing power of these mysterious little cells. They can multiply them by the millions, before they differentiate, creating a pool of blank cells for therapy that won't set off the immune system. But creating a viable, serviceable product from stem cells remains a perplexing task. It seems that for every molecular clue we are given to our biological riddle, we are also handed another vexing problem to solve. And no matter how magical science may seem, what we need is not a magic show—however awesome—but usable products to ease pain and cure disease.

Science is money. If our cells are harnessed and their healing powers understood, big profits can result. The science of biotechnology is a multi-billion-dollar industry, because it has the ability to change the parameters of life itself. How long we live, and how we feel physically and mentally, could one day be improved by biotechnology right before our very eyes. With the human genome now mapped, scientists are able to explore our shared genetic inheritance, and thus our common predisposition to disease. But like any other production, this magic show will come to a screeching halt if one main ingredient is missing: money.

Without cash in the bank, ideas remain dreams. Laboratories remain dark. Medicine stagnates. The reality is that this great dream comes down to dollars and cents. Even though it carries with it the immense hope and responsibility of healing, science is like any other business. It is difficult for most of us to accept that money controls medicine, but life in the laboratory is a matter of trial and error; for months researchers can remain stymied, while the cash drains from their pockets. The life business—like life itself—needs a steady cash flow to survive.

---

Biotechnology has had a perilous financial lifeline in its three decades of existence. In 1971, Cetus Corp. was launched. The science was so new that few understood where its strengths lay, and which technology would generate profits. So Cetus focused on every known biotech platform. It put its money in vaccines, therapeutic proteins, even devised a new method of making antibiotics. Investors quickly became frustrated with the firm's haphazard approach. One of Cetus' disgruntled investors was Bob Swanson, who had made a name for himself at Kleiner, Perkins, Caufield & Byers. This young and ambitious venture capitalist decided to do his own digging to determine what was hot and what was hype.[1]

At the University of California San Francisco campus, Swanson discovered the findings of Herb Boyer and Stanley Cohen, who had brought recombinant DNA—DNA reassembled using components from different sources—to life in the laboratory. Boyer, a University of California biochemist, had discovered a small DNA molecule that could be used by Cohen's gene delivery system—the plasmid. At a 1972 conference in

Hawaii, Cohen, a scientist working out of Stanford, listened in awe to Boyer's seminar on "molecular scissors" that could snip away genes that were then inserted into the DNA of another, completely different organism. A meeting at a Waikiki Beach deli sealed the partnership and led to scientific history. Boyer and Cohen took a gene found in toads and inserted it into a manufactured strain of the bacterium *E. coli*. They got the bacteria to continuously divide and copy the foreign gene. With the speed of the bacteria replicating, genetic engineering was brought to life.[2]

Swanson discovered Boyer and Cohen's findings in the November 1973 issue of the *Proceedings of the National Academy of Sciences*. Then came another informal meeting that would make biotech history. The 27-year-old Swanson met Boyer at Churchill's bar in San Francisco, and subsequently they sealed a deal to create a business based on this novel technology. Researchers and investors could easily have thought that Swanson was rushing the science to market. Recombinant DNA was new and therefore risky. Cohen did say that a saleable product was years in the making.[3] But Swanson wouldn't have any of it. He felt the time to cash in was upon them. And after their meeting, Boyer agreed.

The plan was this: produce proteins from recombinant DNA to be used in disease therapy. The technology was there, as was the ability to mass-produce the product. All they had to do was take a protein pivotal in alleviating the symptoms of a particular disease, transplant it into bacteria, and let it reproduce in that "living factory." Manufacture tonnes of the stuff, harvest it, and *voila*—a product that could be produced consistently and uniformly in mass quantities. They set their sights on manufactured insulin, the life serum of millions of diabetics around the world, and there was hope as well as dollar signs in their eyes. It was a product that could not only generate huge profits, but could also stave off the terrible long-term damage of the disease and save lives. Therein lies the dream of biotech: to do good, and make millions in the process.

Up until this point, insulin was harvested from pigs and cows, which have insulin extremely similar to that of humans. Yet it was still possible for certain diabetics to have allergic reactions to the foreign protein. If the Boyer/Cohen formula worked, the risk of rejection would be eliminated, because human insulin would be produced from the body's own raw materials. Laying down a nominal US$500 apiece to incorporate, Boyer and Swanson sealed the deal, launching the world's first real biotech firm. Swanson proposed the company name "Her-Bob" in honour of the two brains behind the science. But Boyer came up with the catchier "Genetech," short for genetic engineering technology.[4]

Swanson left Kleiner & Perkins to follow his new business plan, which quickly proved to be a good one. Kleiner & Perkins believed so much in its former employee's idea that it sold its shares in Cetus and plunked US$100,000 into Swanson's new venture. Swanson didn't let the excitement get him ahead of himself. Rather, he began what would become a billion-dollar business with baby steps. Genetech needed product rights to generate cash, so he laid out the plans for the company's first research academic partnership virtually, from Kleiner & Perkins offices. He was intent on generating cash before spending it. Genetech's first office was far from big pharma's world of polished dark wood and spotless tinted glass. It had the vibe of a laboratory, with lunches dispensed from a vending machine rather than served in a fancy cafeteria.[5]

Everyone was fuelled by the mind-boggling possibilities—including Swanson, who was known for putting in long hours and asking thought-provoking questions, and was never above doing a job that needed to be done. According to biotech consultant and author Cynthia Robbins-Roth, in order to figure out the environment of a biotech or other high-tech (or even a no-tech) entrepreneurial firm, you have to look at its leaders. They possess the ability to shape the company culture. "Bob Swanson single-handedly drove Genetech into existence, coupling strong business instincts with a love of the science and a willingness to step off the usual CEO pedestal to do whatever it took to keep the company moving forward."[6] Former employee Jim Gower witnessed Swanson's refreshing attitude to business: "Biotech execs need to be able to do whatever it takes. I remember when Bob Swanson was touring some Japanese businessmen through the Genetech halls. In the midst of discussing some high-level project, he stopped to fix a leaky sink and never dropped a beat in the conversation."[7]

While Swanson and the Genetech crew had been driven by scientific excitement for several years, it wasn't until 1980, when Genetech went public, that the first generation of biocapitalists was born. On October 14, 1980, the financial world was introduced to a company that had created the first real biotech product. In 1978, Genetech had sold world rights to recombinant human insulin to Eli Lilly & Co., which won FDA approval for the product in 1982. Buying into the mass-production of human insulin was like striking biotech gold. Even Swanson didn't realize at first what he was sitting on. He was advised to price the stock at US$40, but he felt that might be too high. After all, he was asking people to invest in a company that wouldn't be able to generate a product until 1984. It was such a high-risk venture that the offering memo's cover cautioned, HIGH DEGREE OF RISK.[8] So the stock offering began at US$35. Within 20 minutes of hitting the market the stock soared to US$89, and along with it soared the future of bioentrepreneurs. Genetech set off a biotech frenzy. Science was sexy; science was money. Even if fledgling companies didn't have a product much past the conceptual stage, they were given hope that investors might see the potential. Scientists had long argued that the science was there to bet on. Now investors were ready to listen.

---

Many companies tried to duplicate the success of Genetech, focusing on valuable proteins. Riding on the heels of human insulin was epoetin alpha (EPO), a protein that convinces the body to generate red blood cells—a valuable treatment for people depending on kidney dialysis. So California-based Amgen manufactured recombinant epoetin alpha. Amgen—the heftiest biotech company, worth US$64 billion—now generates US$2 billion annually from this protein.[9]

Investors continued to bet on the biotech dream into the first few months of the nineties. Then people began looking for some return on their investment. Even though the science still held the excitement that had attracted them to invest, they were becoming impatient for the payoff. How long would it take to get a product to market? How many hiccups would they have to endure in the lab before the scientists got it right? By 1993 investors were turning their backs on biotechnology after a series of product failures hit the papers. Suddenly it seemed that biotech would not in fact "improve the odds of product development."[10] Bioentrepreneurs could ride the dream for only so long. As Robbins-Roth explains, these high-powered execs learned that scientific potential could take a company only so far. Investors had to be able to see not only the path to the product, but the product itself. Even if a biotech company had success during early product development, this didn't negate the inherent risk of the industry. Investors understood that things could go wrong during clinical trials, and as late as a Phase III trial, when new treatments are compared to existing ones. It was apparent that research and development (R & D) would attract far less funding if there wasn't substantial stock excitement.[11]

The doubt over whether biotechnology will reduce costs and make money is similar to the doubts in the early days of information technology. During the 1980s and early 1990s, people were dropping massive amounts of cash on computers and software—over 3 percent of the GDP was being spent on the PC craze—even though information technology companies were not reporting an increase in efficiency. According to Nobel prize winner Robert M. Solow, "You can see the computer age everywhere but in the productivity statistics."[12] While it took years to profit from computers, it now seems that virtually every business is reliant on a PC. Biotech, like IT, needs time to show its worth.

It has been estimated that it takes 15 years for a biotech company to generate a profit. Many simply can't survive that long, and the industry is feeling it. The biotech sector as a whole has racked up losses of nearly US$6 billion,[13] which no doubt brings back memories of another cash-poor, idea-rich industry. When the Internet was first introduced, it was believed to be the virtual highway to riches. And in fact people were able to cash in. Unlike their biotech counterparts, which need well-equipped, expensive laboratories as well as trained scientists, Internet companies can launch from a single PC in a dingy garage anywhere in the world. Pubescent-looking entrepreneurs were wowing investors into spending serious amounts of cash. This couldn't be just a phase, many concluded—or at least hoped. But the bubble finally burst when it became apparent that only a select few companies could devise a marketable product based on the technology. Firms like Google, with its astounding August 2004 initial public offering (IPO) of $85 a share—valued at US$1.67 billion—continue to prosper. But for hundreds of non-Googles, the Internet was an insanely expensive bad connection.

Biotech has what a lot of other industries do not: the prospect of changing life as we know it. Sure, the information age changed how we do business. But biotech, like health care, impacts all of humanity. It is not simply a good investment. It is about lengthening and improving life, which is why many investors accept the fact that they must wait decades for their return. People want to believe that biotech will live on, not only to improve their bank accounts, but also to improve humanity. As author Richard W. Oliver says, biotech is everyone's business because it affects us all. Every day, 10,000 more Americans turn 50. By 2025, 30 percent of the U.S. population will be over 55 years of age.[14] Everyone seems to know someone who has developed Alzheimer's, Huntington's, or some other genetic disease.

So when a biotech company proposes a novel drug therapy for, say, Parkinson's or cancer, that could extend and improve lives, people listen—and cross their fingers. As investor columnist Herb Greenberg writes, "With biotech, all it takes to spark a buying frenzy is one major breakthrough in the quest to cure cancer, AIDS, or another disease. And unlike regular technology, which goes through cycles, biotech doesn't create a commodity that can quickly be leapfrogged by a competitor."[15] So even though the global market isn't yet flooded with new biotherapies, its annual worth is estimated to be more than US$30 billion, up from US$12 billion a few years ago.[16]

---

In the 20 years since recombinant insulin made it to market, the industry as a whole has been lagging way behind. But biotech is now getting its goods to market, and many more drugs are in the final stages of development. According to Statistics Canada, 54 percent of Canada's current 18,020 biotech products and processes either are approved, have made it to market, or are in production; the rest are still in development.[17] Biotech is ready for another boom. The leading 100 biotech companies, 70 of which are American, pulled in more than US$21 billion in revenues in 2002, an increase of more than 26 percent from 2001.[18] Yet the harsh reality is that biotechnology is in the same boat as the pharmaceutical industry. It needs lots of time—and money—to show its worth. The average drug for the mass market costs US$350 to US$400 million to develop.[19] The process can take anywhere from a decade to 15 years.

The only way a biotech firm will succeed is by investing in R & D, because without fresh ideas there are no new drugs and there is no constant revenue stream. Recently, however, a radical change coming down the drug pipeline has been driving big pharma into the arms of biotech companies. It's called pharmacogenetics, and it's the first true attempt at personalized medicine. With the human genome mapped, scientists and entrepreneurs alike are hoping that our genes will reveal which drugs cause dangerous side effects in which patients, and why. Maybe one form of a gene predisposes carriers to remain unaffected by a medication, while another form ensures that its carriers will suffer adverse effects. Rather than going through countless years of preclinical and clinical trials, physicians may one day be able to take a drop of a patient's blood and immediately understand his or her drug susceptibilities. This could save millions in drug development, while improving people's health and minimizing deaths. Tailor-made medicine could cause the next boom in biotech. The U.S. genetic testing market alone is estimated to be worth anywhere between US$7 billion and US$20 billion by 2010.[20]

A lot is being invested in biotech/big pharma partnerships on both sides of the microscope. But all this opportunity does not come without risk. According to the pharmaceutical research company Cutting Edge Information, 67 percent of pharmaceutical development deals will fail for reasons that include not being able to meet revenue or scientific goals, poor communication between companies, badly negotiated deals, undefined partnership roles, and poor leadership.[21] Also, the biopharmaceutical industry seems to be running scarily close to pricing itself out of the market. According to Segal Co., a New York-based employee benefits consultancy, the average monthly cost of biopharmaceuticals created from organisms instead of chemicals is US$1,300 per individual, contrasted with US$60 to US$80 per month for traditional drugs. For example, while Zevalin and Rituxan target lymphoma but don't cause the hair loss and exhaustion that tend to come with chemotherapy, these two drugs cost US$28,000 and US$3,530 per dose, respectively.[22] There is a lot of excitement surrounding Enfuvirtide (pentafuside), the biodrug developed to combat the effects of AIDS in those not responding to older AIDS medications. But it takes 106 steps to make Enfuvirtide, compared to the eight to ten steps to make older AIDS meds, and it requires 44 ingredients from 10 countries. This means the therapy rings in at US$20,000 per year.[23] Drug delivery also increases the cost of many biopharmaceuticals. Some must be injected by a health care professional, which is more expensive than popping a pill.

How can biotechnology survive and prosper when the industry is astoundingly more costly and risky than the precarious pharmaceutical market? What does it take for firms to travel the long, dusty road of drug development and approval without losing their investors? Without money, a biotech company will never make it—because in this business, it takes money to make money. Without an investment in the future, there won't be one. Moving from the theoretical arena of the laboratory into the practical arena of the market, with a product to sell, takes a lot of time and a lot of cash.

Canada is a global leader in the biotech laboratory. Revenues reached $3.6 billion in 2001, and the average yearly growth in the sector's revenues has increased 44.7 percent since 1997. The Canadian biotech industry spent $1.3 billion on research and development in 2001, with an average annual growth rate of 28.3 percent, according to Statistics Canada.[24] While there are 470 biotech firms in Canada—the second-largest number worldwide (with the U.S. having the most)—most of these are small, fledgling firms. Two hundred and forty-three have fewer than 10 and 153 have fewer than five people on staff, according to Ernst & Young.[25] In the

United States there are more than 1,450 biotech companies, with more than 300 of these publicly traded and a market cap of US$200 billion. Biotech revenues in 2001 reached US$35 billion, up from US$8 billion in 1992.[26] According to the Biotechnology Industry Organization, annual R&D costs for the U.S. are more than US$15 billion.[27] To understand what it takes to keep a firm alive in this risky, complex industry, just look at an American biotech legend: Geron Corp.

---

In high school, Michael West found God in his hometown of Niles, Michigan. His grandfather had recently passed away, and for the first time in their lives he and his father began attending church. West became spellbound by the possibility of immortality. He thought of the earth dying every fall and returning to life each spring, and learned that the rebirth of nature was God's way of telling him he would be resurrected after death. "The bible absorbed me as a sponge absorbs a drop of water," West writes in his book, *The Immortal Cell*. "What a glorious solution to the problem of mortality. This, I realized, was the greatest force people feel drawing them to religion. It was a magnificent story of love conquering death."[28]

The only thing in West's life that could compete with this devoutness was his devotion to science. West is tall and fit, with dark brown hair and a clean shave. He is often dressed in a shirt and tie peeking out of a crisp white lab coat. His expression is mild-mannered, yet his eyes reveal a dogged determination that he seems to have been born with. When West began his master's degree in biology at Andrews University, a Seventh-Day Adventist school in Berrien Springs, Michigan, he decided to put his scientific skills to use and prove the theological view of life. He was studying the history of fossils, and was intent on harmonizing it with the Bible. But as much as he tried to convince himself that the Good Book's views on the beginning of life were correct, his scientific mind convinced him otherwise. "There was absolutely no conceivable way that I could hang any hope any more on a biblical view of the origins of life on earth. I mean, Darwin was right. He was absolutely stone-cold right."[29]

Rather than give up on immortality, West turned to science to give him new life. One day he found himself sitting across from the cemetery where his father was now buried alongside his grandfather. "It was like Buddha under the tree. Something welled up within me and what I saw at the cemetery was not just the death of my father. . . . All the people I loved and cared for that were still alive would have their names etched in those same tombstones," he says. "It's certain, and the day will come. And when I contemplated that, something dramatic happened to me."[30] He became overwhelmed by a hatred for death. And with this, his life mission was decided: to study the natural process of aging—and challenge death.

He began by examining human cells, studying the process of cellular aging. How do cells proliferate and repair themselves inside the body? What makes them live, and what makes them die? He turned to the experiments of Leonard Hayflick, who in the 1950s was questioning the immortality of cells. Hayflick's experiments led him to believe that, contrary to prevailing scientific thought, normal cells might actually be mortal. The only immortal cells that Hayflick witnessed appeared to be abnormal. Something within them wasn't right; they might be cancerous, or they might be missing a chromosome. Hayflick postulated that cells were naturally mortal, and only exhibited extended life when they were hurt genetically.[31] When he cultured normal cells, he could make them divide only 50 times, contrary to the insistence of his colleagues that cells continued to divide.

Hayflick decided to settle the question by proving that there was a limit to the lifespan of normal cells. His method would be to place young and old cells within the same Petri dish. If normal cells did have an aging clock, then the young cells would continue to divide after the old cells stopped. To differentiate between the two, Hayflick used young female cells and old male cells; he dubbed the procedure the "Dirty Old Man Experiment." Much to his excitement, the experiment worked. The young cells kept dividing after their old mates had stopped. To verify his findings, Hayflick then performed the "Dirty Old Woman Experiment," using young male cells and old female cells. This experiment was also a success, and so was Hayflick. Today, cellular aging is explained with reference to the "Hayflick phenomenon" or "Hayflick limit."

When West discovered Hayflick's work, his mind reeled. Here was the evidence he was looking for to begin experimenting with aging. West was interested in telomeres—the ends of chromosomes, named by scientist Hermann Müller, who studied their fascinating abilities. "Telomere" is derived from two Greek words, *telos,* "end," and *meros,* "parts." Telomeres have been compared to the hard plastic ends of shoelaces, because they perform the process of "capping" chromosomal ends to indicate whether the DNA has become damaged. If there is no break in the telomeres, the DNA is intact.[32]

West was specifically intrigued by the phenomenon of telomeres shortening. This is a natural part of the life cycle of a cell. During cell division, DNA makes a copy of itself. But the DNA at the end of each chromosome can't be duplicated. This DNA is the telomere. Once telomeres shorten to a certain point, the cells they call home lose the ability to

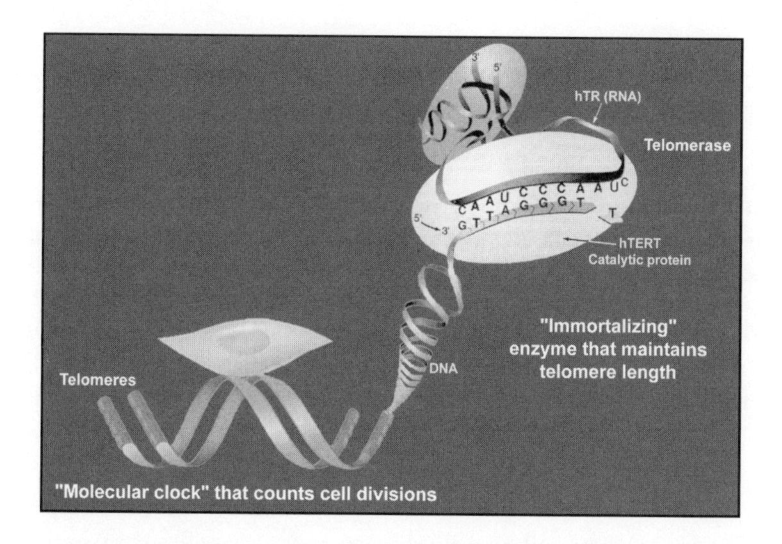


**The human telomerase enzyme extends the telomere of a chromosome. Geron and its collaborators cloned the key catalytic protein (hTERT) and demonstrated that telomerase extends or maintains telomerase in human cells.**

—Geron Corporation

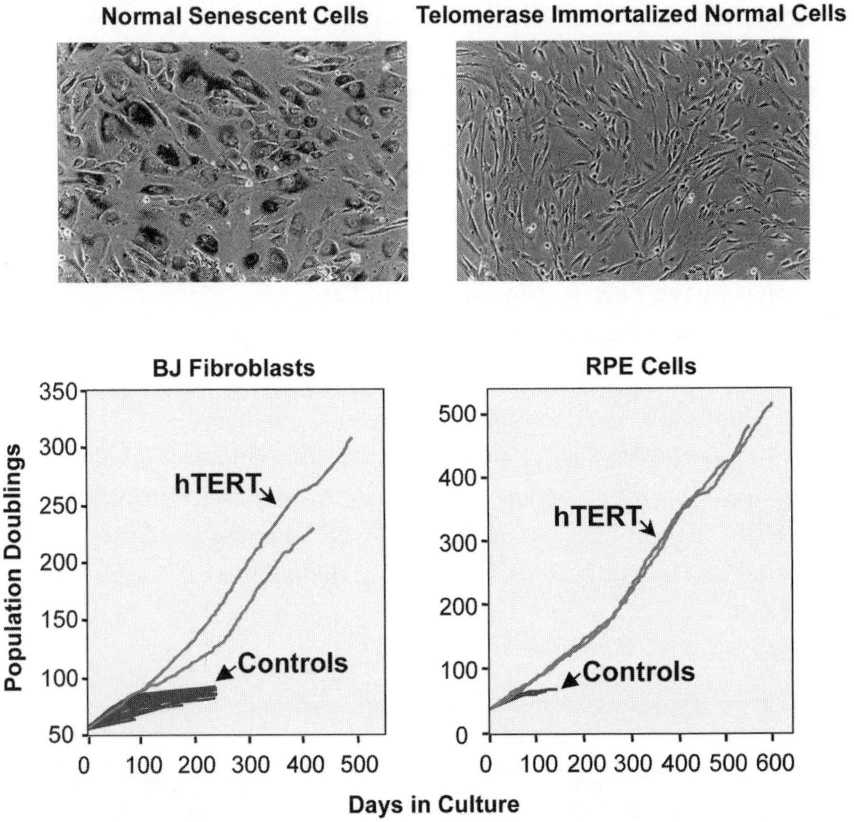

Normal Senescent Cells

Telomerase Immortalized Normal Cells

BJ Fibroblasts

RPE Cells

Population Doublings

hTERT

Controls

Days in Culture

**Telomerase was introduced into normal human cells; control cells did not receive telomerase. Control cells underwent normal cell divisions for aging. The telomerase immortalized cells underwent even more divisions. Both were tested for a biomarker associated with aging. The top left image of cells exhibited the biomarker strongly, while the other telomerase immortalized cells showed virtually no biomarker. The graphs show that the telomerase enzyme (hTERT) increases the lifespan of a cell.** —Geron Corporation

divide. The cells begin to age, and eventually die. All because of the power of its telomeres. To study this process more closely, West looked at the work of several scientists and was thrilled by what he discovered. Hayflick and a fellow scientist named Woody Wright had revealed that the aging clock of a cell could be found in its nucleus. Yale University's

Gregg Morin had discovered telomerase, the enzyme responsible for extending the telomeres. Carol Greider, from James Watson's laboratory at Cold Spring Harbor, Bruce Futcher, and Cal Harley, at McMaster University, had witnessed telomeres shortening in aging cells in vitro. And finally, West learned, Howard Cook had discovered telomeres shortening in cells that were aging normally.[33]

West had come up with the first crucial ingredient needed to launch a biotech firm: the idea. If telomeres could be harnessed and understood, if telomeres could be tricked into retaining their length rather than shortening, then cells could stop aging. If normal cells were taught to continue dividing, they could replicate indefinitely, like cancer cells. Harnessing the power of telomeres might in fact stop such diseases from taking over the normal cells of the body, and maybe immortality itself would be within humanity's grasp. You don't get much better intellectual property than that. Now into his medical degree, West had to make a choice: "write grants, assemble maybe a dozen lab personnel if I was very lucky, and then start to work on the effort I had in mind. . . . Alternatively, I thought, I could try to launch a biotechnology company and eventually, over a three- to five-year period, end up employing perhaps 120 people—far more, needless to say, than would fit into the cramped quarters of a medical school lab."[34]

West decided that a biotech firm would have more steam to drive their new endeavour than a small academic lab. Now he needed a name for his company. He turned to the New Testament for the name Geron.[35] *Geron* is Greek for "old man"; it's what gerontology, the study of aging, gets its name from. Woody Wright and Jerry Shay joined Geron's scientific advisory board, and the scientists got busy filing patent applications so they could protect their experiments. West and his team thought of every conceivable use for telomerase (an enzyme that controls the replication of telomeres), from attacking skin aging and macular degeneration (which causes vision loss) to combating AIDS and hardening of the arteries. All of these conditions weaken the body by shortening the lifespan of cells.

While West's instincts told him that telomerase could one day be translated into important products, he was far from making this a reality. He still needed to track down the precious telomerase gene, and he needed to find the money to do it. "There were, of course, a few major problems in trying to launch the company," he admits. "I couldn't actually prove that all of this was real—that *any* of it was real (not exactly a minor point when asking investors to cough up hundreds of thousands of dollars in seed capital). Instead, I had to convince them to believe in my instincts. And I had only vague ideas of how we were going to track down the telomerase gene."[36]

How, West wondered, could he convince investors to bet on an entrepreneur as green as himself? West needed to find some "angels," investors who seem to come from heaven above. The term "angel investor" was first used to describe New York investors who would bet their money on Broadway, well before the plays could generate a profit. Today, angels often lend start-ups their wings to get off the ground. According to the University of Windsor's former dean of business Rein Peterson and professor Joel Shulman, 95 percent of all start-ups in countries within the OECD were financed by angels and "love money" from families of the founders.[37] In Canada alone, 15 percent of the capital raised by biotech firms in 2001 came from angel investors.[38]

West's first prospective angel turned out to be definitely of this earth. He was a wealthy man in Santa Barbara, California, who was frustrated by his own process of aging. West went to the man's office to make a slideshow presentation. A few minutes into it, he glanced over at the man, whose head was drooping dangerously towards slumber. Soon he was fast asleep. Devastated, West packed up his things and left the man's office, with every intention of returning to medical school and forgetting this crazy biotech dream. But it wasn't long before West got the phone call that would change his life. Dr. Bill Ryan, a biotech analyst at Smith Barney, asked him to present his ideas. And when he did, West was bowled over by Ryan's response: "You know, this could be the greatest revolution in the history of biotechnology!"[39] West was given his first fat investment cheque, and Geron was launched.

Another endearing angel is a retired oilman from Houston who, as president of the Curing Old Age Disease Society, is intent on finding the fountain of youth. Born in 1914, Miller Quarles wards off the ravages of age by taking 50 pills a day, which includes an array of food supplements and vitamins.[40] Quarles and some of his oil buddies were intrigued by West's business plan, and decided to invest. West was able to raise US$250,000, with Quarles putting down US$50,000 himself. (The average amount of seed money it takes to start a biotech is between US$250,000 and US$5 million, yet numerous companies attempt to launch with much less.[41]) This gave West the funds to set up an office, and to continue the search for deep-pocketed investors who would make Geron a real player.

In November 1991 West got his next big break. He was invited to speak at the National Conference on Biotechnology Ventures, a meeting of venture capitalists in Silicon Valley. "As luck has it—you know we all have good and bad hair days?" he says. "I was really fired up that day." He explained his remedy for an aging society to the audience, and a group of hungry investors took notice. After the presentation, he was gathering up his slides when a group from Kleiner, Perkins, Caufield, & Byers approached. "They literally formed a circle around me. And they walked me, surrounding me, into a little room and closed the door."[42] KPCB wanted to lead the financing. A few months later, West sealed the first round for Geron: US$7.5 million. Forget penny-pinching academia. West had entered the big leagues.

The way to study telomeres, West decided, was to experiment on embryonic stem cells that had not yet differentiated. In the mid-nineties, stem cell research had stagnated in academia because of President Clinton's legislation cutting federal funding for such controversial research (see Chapter Six). This didn't affect Geron's progress, since it was a private corporation. West experimented with human fetal testes, but his cells refused to proliferate. He soon discovered that they were too old; his experiments required five-week-old fetuses. A group of academic scientists had such materials, but their own research was stymied for lack of federal funds.

West flew out to the University of Wisconsin–Madison to meet James Thomson, who was going to publish a paper in a scientific journal on his isolation of embryonic stem cells from a certain type of monkey. If West could get his hands on these cells, or—even better—on those cells in monkeys as well as all primates, including humans, his telomerase research would again be up and running. But while Geron agreed to provide the money to fuel Thomson's research, the university refused to grant the company exclusive rights to the work. Instead, West was given access to Thomson's cells, which enabled Geron to continue its study of telomerase. It was soon apparent to West that he was studying the true cell of longevity. And it fascinated him beyond belief. He knew that if they could coax these cells to multiply by the millions, they could form new therapies to cure life's most vexing diseases. "They were life itself, the immortal thread that connected the generations," writes West. "And they were the *fons et origo* of all the cells in the body." Think how happy this would make Geron's investors, West thought. He did attempt to enamour his investors with a video of cells beating in a dish, but there was still no practical application. How could these stem cells be harvested, and their powers put to medical use? Geron had to return to the laboratory until its scientists could grow human embryonic stem cells, showing that they could in fact be isolated and cultivated.

In 1996, Geron decided to raise some much-needed cash through an IPO. When the company went public, West says, "It was all dreams."[43] That year, several IPOs were close to the US$100 million mark—which handed biotech legitimacy in the world of business, and told investors that it was the next area to watch.[44] When a market is hot and investors want in on the action, companies have to profit from the excitement while they can; as history shows, it won't last forever. Geron decided to cash in on the enthusiasm surrounding telomerase as a novel therapy that could change the management of cancer. Geron had also begun work on embryonic stem cells that would later put the company in science's history books. It was in the position to make some money by entering the public market. So West and two of his colleagues boarded a private jet for a non-stop tour of the United States and Europe, generating investor interest. Geron finally opened on the Nasdaq at US$8 a share. For fiscal 1996, the company reported revenues of US$5.3 million and a net loss of US$10.7 million.

---

Generating investor excitement in biotech is an arduous task. Companies can no longer rely on the future miracles of regenerative science alone. Investors need to understand how an idea will evolve into a saleable product, and it is often hard for them to feel confident that a certain technology will show a return when so many other firms are trying to cash in on the same platform. That is why product-based companies often do better than technology-based companies. Drugs are also written up and critiqued in scientific journals, which offers the company exposure as well as credibility.[45]

Taking a company public means there has to be more accountability. Every action that could be of importance to investors must be publicly disclosed, and if there is a setback, no matter how insignificant, it must be reported. This becomes a problem when investors do not understand the intricate technology, and may become overly worried when in fact there is a simple explanation for the setback. Along with all the heightened disclosure and securities reporting comes a loss of control over business decisions, pressure for constant company growth, and, most of all, cost: the average underwriter takes 6 percent of what is raised by an IPO, not to mention marketing, auditing, printing, and legal fees.[46]

Like other public companies, biotechs find ways to manage public disclosures. Some choose to deliver good news as it happens while delaying the bad—say, releasing it after the markets have closed on a Friday, when the press is less likely to pick up on it. Others distract investors from traditional earnings per share by focusing on pro forma earnings. Capital markets react more strongly to non-financial disclosures—such as progress in the drug development process—than to earnings announcements. So any product advancement—a move from the Petri dish to animal trials, an alliance with a big industry player, or the beginning of clinical studies—is likely to be well publicized, to divert attention from the company's stock price.[47]

Of course, there are also many advantages to going public. The firm is handed a huge amount of capital to keep the lights on and the microscopes staffed. It can use stock rather than cash to form a strategic alliance with a big player. An IPO heightens public awareness of the firm. And liquidity is generated for shareholders.

After an IPO, the firm is quickly called on to show investors how far it has progressed with their money. What has happened since the offering? What have the scientists discovered, and how will this bring the market closer to reality? Biotech's aim is getting a product into the hands of patients who can attest to its importance, thus increasing the company's worth. But the drug approval process is painfully slow. While Health Canada aims to review a drug in 18 months, this goal is not always met. And this review must come before clinical trials are allowed to start, pushing the market further into the future. Even after a Phase III trial has been completed, it takes two more years to gain approval from the regulator. In the U.S., President George W. Bush has taken notice of the delay, fuelling the FDA with US$1.4 billion, some of which is designated to help speed up the approval process—a process that already takes only half the time it takes in Canada.[48]

Every day that is lost to review and approval increases the risk that important science might be pushed by the wayside, says Janet Lambert, president of BIOTECanada, the country's biotech industry association. The average Canadian company has no more than 18 months of funding in the bank, she says, adding that it takes Health Canada about 800 days (about 26 months) to work through the approval process. "It's not for lack of good science that we're seeing companies go under. It's for the time it takes," says Lambert. "It takes a year before Health Canada even picks up a submission from the shelf. We call it dust time."[49]

According to an Ernst & Young 2004 report on biotech, there were 81 Canadian public companies in 2003, a decrease of four since the previous year. There were no IPOs in biotech that year. Revenues from public companies increased by 18 percent in 2003 to $1.729 billion. Private firms, on the other hand, increased by 17 percent, to 389. Even though there are 470 biotech firms in Canada, the top 10 represent 70 percent of the total market cap for the entire industry, raising more than $773 million.[50] For the small biotechs to survive, they typically must form an alliance with a big pharmaceutical company seasoned in the risky and costly drug development process. Such a partnership offers cachet in the eyes of investors, as well as money in the bank.

Interestingly enough, big pharma is just as dependent on such alliances. In 2003, analysts announced that the pharmaceutical industry was running out of breath in terms of productivity and return on investment. While more money is being spent on R & D, there has been a decrease in the number of drugs that actually get into the hands of patients.[51] Also, many of the blockbuster drugs that were developed by pharmaceutical giants are losing their patent protection. Generic drugs (cheaper copies of brand-name drugs that are sold once patents expire) not only offer customers choice, they can also loosen big pharma's monopoly on certain medications. As the medical world moves increasingly towards cellular and genetic therapies, the drug companies must invest in biotech to stay in the game.

The average strategic alliance between biotech and big pharma usually goes something like this. Big pharma pays a licensing fee for rights to use the biotech's technology. The pharmaceutical company provides the biotech with R&D funding for the span of the agreement, typically an initial three to five years, so the biotech can afford staff and other costs of its research. The biotech is then rewarded by its partner with payments as it reaches certain milestones—starting clinical trials, receiving government approval—that advance the product towards the market. Big pharma sometimes purchases equity in the biotech, but not always, as it is not always viewed as an asset.[52]

According to management consulting firm SECOR, Canadian biotech companies had more than 120 corporate alliances in 2001, mostly with American firms. And 30 percent of biotechs in Ontario, Quebec, British Columbia, and Alberta have alliances.[53] In the United States, pharmaceutical giant Merck & Co. intended to more than double its alliances to around 80 in 2004, from around 30 in 2003. According to American life sciences merchant bank Burrill & Co., alliances were worth around US$2.1 billion for the fourth quarter of 2003.[54]

---

Geron had gained credibility through its alliances with Jamie Thomson and John Gearhart, but it was still searching for something—be it a product or a technology—to profit from. In order to keep investors happy, Geron had to get back to the bench to come up with the science that would deliver future earnings. Only R & D will ensure that valuable products are always "in the pipeline" and on the way to market. Since some drugs take longer than others to be approved, while others fail along the way, companies need a number of products in development at the same time.

West's first and most crucial step in R&D was to clone the telomerase gene—and fast—before the competition could beat him to it. The Geron team had already shown that telomerase was active in cancer cells. They studied cancerous cells from the cervix, kidney, prostate, lung, and skin, and found that 98 out of 100 tumour cell lines were telomerase-positive—that is, the cells had the enzyme they needed to keep growing. None of the 22 normal cell lines, on the other hand, exhibited active telomerase. They knew that around half of all types of cancer expressed a mutation of gene p53, whose function is to throw on the brakes in normal cells to control growth. According to West, half of all tumour cells had a broken p53 brake, which could lead them to grow uncontrollably. Jerry Shay and a team of scientists compared cancerous breast tissue with normal breast tissue, and discovered that none of the normal tissue showed signs of the "immortality gene." They tested 12 tumour types in all.[55]

The results were staggering. If scientists could target telomerase, triggering it to begin the natural aging process of cells, then maybe its powers could be harnessed to kill cancer cells while leaving normal cells intact. Rather than enduring months of radiation and chemotherapy, patients might one day be able to reset their own telomerase and rid themselves of one of the world's most feared killers. The results were published in *Science*'s December 23, 1994 issue. But Geron investors were getting impatient. Other teams of scientists in competing laboratories were also searching for the telomerase gene. Sure, the power of telomerase was impressive. But would Geron be the one to cash in on it? To keep its investors, Geron desperately needed to find it first. West and his team calculated that telomerase was made up of two parts: the "hand" of the enzyme, which made the telomeric sequence of bases, and a thread of RNA (ribonucleic acid) that told the hand to do its job. While both DNA and RNA carry a cell's genetic data, it is the job of RNA to take the genetic information stored in DNA and use it to make proteins. Geron wondered if telomerase would be a reverse transcriptase, a molecule usually residing in viruses that creates DNA from "instructions" handed down in an RNA molecule.[56]

One of Geron's scientists, Junli Feng, found a piece of RNA that looked to be the size expected for the RNA component of telomerase. West took the sample and tested it against a library of RNAs from both normal and tumour cells. The study showed that this RNA was much more active in tumour cells. The RNA was then removed from the tumour cells, to see if the telomeres would act as they did within normal cells and shorten naturally, stopping the growth. *Eureka!* The typically immortal cancer cells were suddenly aging and dying. But while Geron had shown that tumour cells could lose their immortality through the elimination of telomerase, they had not yet proven that the life of normal cells could be extended by giving them a dose of telomerase.

Meanwhile, big things were happening outside the Geron laboratories. University of Colorado student Joachim Lingner isolated and purified telomerase from the pond-water animal *Euplotes*. The protein he cornered, coined p123, looked nothing like the one Geron's scientists had been focusing on. To add more pressure on the Geron team, Amgen proclaimed that it had cloned a component of human telomerase. This rang warning bells for Geron investors, but it was soon discovered that Amgen had simply found a protein related to telomerase, and was no farther ahead in the race to capture the gene. Luckily for Geron, it was able to secure the exclusive licence to the University of Colorado's patent, stepping up the pace to find the human telomerase gene.

Thanks to the Human Genome Project, new genes were continually being discovered and published. Everyone at Geron was given instructions to keep scanning the Internet for a human match to the telomerase of the pond animal. One night, graduate student Toru Nakamura punched in the *Euplotes* sequence. Suddenly a sequence popped up on his screen. It was from an inflamed human tonsil donated by the University of Washington at St. Louis Medical School. Its function was unknown, so it was called Homo sapiens cDNA clone 712562. (Later it was renamed hTERT, which stands for "human telomerase reverse transcriptase.") They had found their gene. But they still had to write up and publish the results before anyone else did. In the summer of 1997, with the competition closing in on the gene, West and his team worked day and night, sleeping in a trailer in Geron's parking lot, to clone the gene encoding telomerase.

On August 15, 1997, their paper was published in *Science*; Geron had crossed the finish line first. On the day of publication, Geron's stock shot up 115 percent. West appeared on CNBC. Geron investors were elated. The company was awarded the coveted patent, giving it exclusive rights to the gene. Now its scientists could test telomerase within normal cells to determine if this protein did in fact give them immortality. The first cells tested belonged to none other than Len Hayflick. While he was visiting Geron one day, he pulled up his pant leg, swabbed a section of his calf, and used a scalpel to lift off a piece of skin. The skin was then cultured, with West joking that he could truly test the "Hayflick limit." West inserted the telomerase DNA into Hayflick's cells. Four weeks later, the results were in. Hayflick's sample of cells tested positive for telomerase. After many different types of samples had been tested, the experiment was declared a success. Geron investors didn't seem bothered that a product was not yet in the works. The potential of telomerase would give West and his team some time to catch their breath.

---

In 1998, Jamie Thomson and his team isolated embryonic stem cells, while John Gearhart's group isolated embryonic germ cells (see Chapter Four). For Michael West, it was an immortal chapter in the book of life. But the people at Geron were looking in a different direction. Instead of focusing on cellular aging, they wanted to create new therapies for disease: to study ways to prevent cancer, diabetes, and heart disease. West quickly made a decision. "I'm normally not rash," he says. "But in this case I knew what I was doing." He would leave the company he had founded in order to pursue his lifelong mission of conquering death. For West, the battle was far from over. "I've always had this criticism that Mike's always out, chasing some new dream," he says. This time, it was cloning. West and his wife, Karen Chapman, one of the leading scientists on the telomerase gene hunt, flew to Australia to sit in on an important meeting on genetically engineering cells to create clones of animals. West and Chapman discovered that cloning could reprogram the internal clock in cells, and recast an animal's lifespan. If normal cells could be reversed to their embryonic state, they could also be programmed to become any type of cell. This would not only provide material for cellular therapies, but also prevent rejection, because patients could be treated

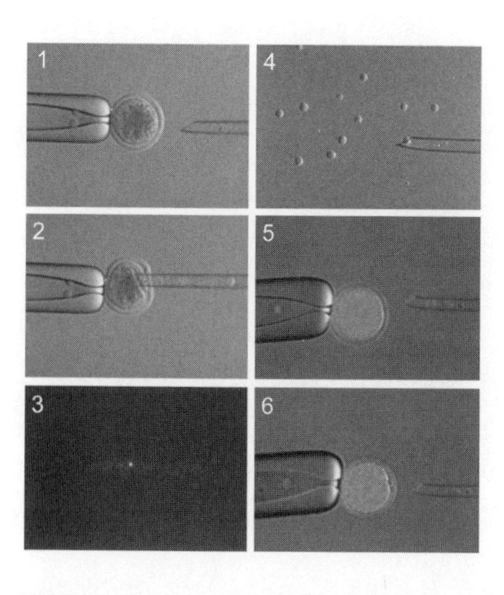

**First example in history of**

- Mammal cloned from adult somatic cell
- Reprogramming of adult mammalian cells

**Improvements**

- Process has been exemplified in sheep, cows, goats, pigs, mice
- Cells have been engineered and then subjected to nuclear transfer

**Nuclear transfer generates human cells or entire animals whose genetic material is derived solely from the nucleus of a single cell obtained from a single individual. The nucleus containing all the chromosomal DNA is removed from an egg cell (see images 1, 2, 3) and replaced with the nucleus of a donor somatic cell (see images 4, 5, 6). The egg cell is activated and transferred to a surrogate mother animal and allowed to develop to term. The offspring is a genetic clone. Nuclear transfer was used to create Dolly, the first cloned mammal.** –Geron Corporation with their own cells. If the short telomeres of an old cell could be reset and stretched to the length of those in an embryonic cell, aging could in fact be reversed.[57]

At about the same time that West decided to leave Geron, Advanced Cell Technology (ACT), a fledgling biotech out of Worcester, Mass., was making a name for itself. ACT was owned by Avian Farms, which made its money by cloning chickens. During a meeting at ACT's offices, West learned what ACT was doing. Scientist Jose Cibelli had taken some cells from the inside of his own cheek, isolated the live ones, and combined them with the egg cells of a cow. His experiment had controversy written all over it. In 1996, Edinburgh's Roslin Institute garnered worldwide attention by cloning Dolly the sheep. Washington was adamant that nuclear transfer should not be performed using human cells; the White House had no intention of letting cloned people run around the country. But ACT was adamant about advancing science. West agreed, and joined ACT as its CEO in October 1998.

While West was searching for immortality at ACT, a new CEO was pumping life back into Geron. After many years in academia—"I had bigger ideas than what I could get funded by NIH"—Dr. Thomas Okarma had founded Applied Immune Sciences, Inc. He joined Geron in December 1997, as head of research, bringing with him a new business plan. While for years Geron had been branded as an anti-aging company, Okarma quickly ended that. "I'm not a person who believes that anybody ought to live to 150 years of age," he says. "Even if you did it right, the problem with the population dynamics would be ridiculous." Not to mention the cost of immortality. "Who's going to have the money to rebuild your brain, your heart, and your kidney?" asks Okarma. "No one's going to rebuild my heart if I'm 95 years old and I've got Alzheimer's."

Okarma, who became CEO of Geron in 1999, has other plans for telomerase. His mission is to improve life rather than extend it. He wants people to live functional, independent lives, and his eye is on products that will meet that goal. When West got Geron's initial venture capital funding, telomerase was new and its prospects were unbelievably exciting. "But what's the product here?" asks Okarma. "How do you translate

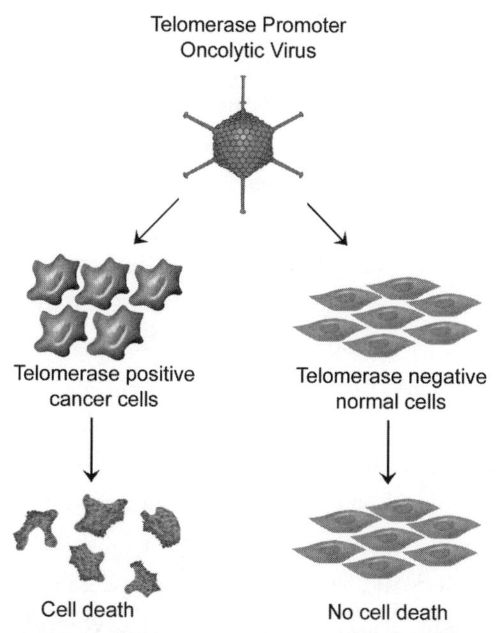


**Telomerase is present in most cancer cells. Customized adenoviruses (common cold viruses) can be engineered to contain the hTERT promoter, thereby causing the virus to infect and kill cancer cells that express telomerase without harming normal cells that don't express telomerase.** —Geron Corporation

that wish into something that's real and returns value to shareholders and is valuable to medicine?"[58]

Geron is now experimenting with telomerase in many types of cells, hoping to generate several important medical advances. Immune cells could be regenerated for AIDS patients, islet cells for diabetics, and retinal cells for the blind. If normal cells can be coaxed into multiplying endlessly without becoming cancerous, Geron is sitting on a gold mine. But the real money-maker, which every day seems closer to the clinic, is cancer therapy. Geron has created two drugs for cancer patients that are designed to inhibit telomerase. GRN163 and GRN163L have been tested successfully in animal models. Mice with myeloma, a malignant tumour that attacks the bone marrow in a similar way in mice and humans, were given GRN163, which decreased their tumours while increasing the survival rate in comparison to the maximum dose of the chemotherapy agent that has been most successful at treating such cancer. GRN163L showed even better results. Both drugs work by binding to the telomerase in cancer cells and inhibiting it without infringing on the function of surrounding normal cells. These drugs have also been tested in other cancers, including cancer of the cervix, the prostate, and the liver, with similarly positive results.[59]

No tumour they have tested is able to escape death by these drugs, says Okarma, adding that five different human tumours have succumbed to this type of therapy. The drug is intended to be administered two to three times a week, as an outpatient treatment, in combination with chemotherapy or radiation, to treat every type of cancer. "As such, that's a mega brand," says Okarma. "And we own it." For example, Geron has reached the clinical trial stage with its telomerase vaccine that targets metastatic (spreading) prostate cancer. The trial, which is happening at Duke University Medical Center, aims to include 24 participants with this type of cancer, divided into a high-dose and a low-dose group. Eighteen patients have been treated, none of whom has experienced adverse effects.

The vaccine aims to "teach" a patient's immune system to attack cancer cells while leaving the other cells alone. The person's immune system is given an antigen—something that causes the immune system to produce antibodies to fight against it. This particular antigen is as similar to cancer cells as possible, to confuse the immune system into attacking the cancer. A kind of T-cell is created to attack cancer cells with active telomerase while leaving normal cells alone.[60] Geron has secured more than 200 patents and patent applications based on telomerase.

Now that Geron knows its commercial target, it must generate viable products. Before launching the cancer therapies, Okarma is rolling out embryonic stem cell therapy for spinal cord patients. Studies have been done on rats with injuries known as thoracic spinal cord contusion. Geron scientists have derived oligodendroglial progenitors, which help form the central nervous system, from human embryonic stem cells, and implanted them into the site of injury. The result has been improved function within one month.

Even though the market for spinal cord injury therapy is relatively small, meaning that Geron cannot expect huge revenues, Okarma is insistent that this be the first product to market. "The purpose isn't to create a mega brand with the first cell type," he says. "It's to demonstrate dramatically, in a patient population that has no alternative, that these cells are literally life-saving and function-restoring. That's going to make a lot of people's eyes open very wide." Once the initial hype surrounding stem cells died down, many investors began wondering if there was in fact something to sell. But a product is exactly what Okarma intends to deliver. And after the spinal cord therapy, he will roll out several other regenerative therapies to maintain investor interest. "What you're asking an investor to do is imagine a world in medicine and therapeutics that is so different from our world today," he says, "that it's hard for them to make that leap in the period of time that we're trying to convince them it's going to happen in."

In the end, the success of a drug comes down to the cost-benefit ratio. Is it likely to generate enough profit and medical benefit to justify the cost of its production? Even if a drug makes it through all three phases of clinical trials, showing that it is safe, effective, and works better than existing treatments, the question still remains: Is it worth actually taking the drug to market, and solving the problems of producing highly regulated therapies both cheaply and uniformly, in mass quantities?

It is estimated that pre-clinical research for a new medical compound costs between US$0.2 and US$0.8 million. A Phase I clinical trial is estimated to cost anywhere from US$5 to US$10 million, Phase II US$10 to US$50 million, and Phase III US$50 to US$200 million.[61] Marketing new drugs also drains bank accounts. After the many years it takes to develop a drug, firms must aggressively market their products if they are to profit from them before their patent protection, and their monopoly, runs out, which is typically after 20 years.

To manage soaring costs, Geron was forced to cut its workforce dramatically. After two rounds of layoffs in June 2002 and January 2003, the company was reduced by two-thirds its size. "It was hard for everybody," says Okarma. "Every time you walk by an empty cubicle you kind of get sick." But investors seemed impressed by Geron's take-action attitude;

the company had decided it was financially prudent to lay off staff and contract out the manufacturing of many drugs. "We did the hard thing," says Okarma. "We stopped some programs, we exported others to academic collaborators, but we didn't lose any momentum on the important things to demonstrate progress."[62]

ACT, on the other hand, remains a maverick lab. With 18 staff and some big ideas, West is again working to raise capital. To do so, he continues a dangerous courtship with the media, giving journalists access to his work without truly knowing how he will be portrayed. There was the time he allowed the television news program *48 Hours* into ACT's laboratories. It was "sweeps week," and the newsmagazine was looking for a story that would keep viewers riveted. Hoping to generate investor interest in his new company, West took the TV crew through the process of how DNA was removed from a cow egg cell and transplanted into a blank human egg cell. He assured reporters that they weren't creating half cow/half humans, but working to create reborn cells that could be used in many different medical therapies. Knowing that such controversial work could be misconstrued, he told *48 Hours,* "The lives of thousands of sick people lie in your hands now. . . . You need to take what you've just filmed with the utmost seriousness." Once the crew had left, West felt something hard crack under his shoe. It was eggshell. The journalists had done an experiment of their own; they had taken a hard-boiled egg, dipped it in liquid nitrogen to freeze it, and kept the cameras rolling as it fell to the ground. "Did they explain why they would do such a crazy thing?" West asked one of his technicians. "Something about how the use of human eggs in cloning could shatter the fabric of society," he was told.[63]

West continued to compete with the company he had founded. In May 1999, Geron acquired Roslin Bio-Med, which owns the licences to the Roslin Institute's patents, including Dolly the sheep. It was Okarma's team, not West's, that was able to announce that Dolly's telomeres were prematurely short, and that the famous sheep, born from the cell of a six-year-old ewe, had been born old. When Dolly reached three years, it was found that her aging clock had not been reset, and that she was actually nine years old. ACT made its own headlines when it proclaimed that it had cloned human embryos. West's announcement that ACT had successfully created stem cells from cloned embryos was published in the November 25, 2001 issue of *e-biomed: The Journal of Regenerative Medicine,* despite the fact that four of the on-line journal's editorial board resigned in protest. Among them was John Gearhart, whom West had personally approached to join Geron years earlier. Many scientists were outraged by the news. They said West was simply grabbing media attention in order to raise some venture capital for his firm.

The science was no good, said some big names in the field, including MIT cloning legend Rudolf Jaenisch: "I have major problems with this ludicrous, outrageous, failed experiment." Jaenisch explained that ACT had started with 71 human eggs, with only three able to divide at all. Then these three eggs had died shortly after. The results, he said, amounted to "third-rate science."[64] Cell biologist Jerry Shay, who met West when he was still a medical student, says West "is always trying to hype his position." West maintains he would never publish scientific findings just to raise capital, yet he does concede that ACT needs a lot more cash in order to survive.[65] "I have a thick skin, but it hurts," he admits, referring to the criticism. "There are days that I go home and think, 'Why don't I just take a university position?' But I keep coming back to the fact that I know we can help people."[66] Despite the opposition, the devout boy from Niles, Michigan, has never lost sight of his mission. He will continue in his quest for immortality. "I push, push, push," he says. "And when you push fast, hard, you offend people . . . if we're going to get anything significant done, we've got to go so much faster than we possibly can. So that makes us labelled as being reckless."[67]

In the beginning, when Michael West first decided to study telomerase, his wife, Karen Chapman, asked him if he truly believed that it could extend life. West said yes, he did believe it. "Yeah right," she contested. "When pigs fly." On West's desk there now sits a pig with wings.[68]

## Chapter Eight

# Owning Ourselves (and Each Other)

## The Patenting of Life

In 1980, an Indian microbiologist convinced the world that a living organism could be owned, bought, and sold. Despite being hailed as sacrosanct and divine, life, it seemed, was in fact a commodity to be patented, and profited from. This microbiologist challenged the belief that life could only be engineered by God or nature. Instead, scientists were the real "architects of life."[1] Ananda Mohan Chakrabarty will forever be famous for patenting a bacterium that he engineered to devour oil spills.

In the late 1960s and early 1970s, oil spills became a grave concern, as coasts around the world were left blackened and sticky when oil escaped from tankers. Chakrabarty, a patient man with a wide, infectious smile, worked as a scientist for General Electric in Schenectady, New York. He knew of a family of bacteria called *Pseudomonas* that target certain hydrocarbons found in oil. These bacteria contain plasmids that "eat," or break down, oil. There were many types of *Pseudomonas* that could nibble away at hydrocarbons. Each type attacked different components of crude oil, but none was strong enough to consume a massive oil slick.[2]

Chakrabarty knew that if a mix of *Pseudomonas* could be created, many hydrocarbons could be broken down. But the process wasn't that simple. According to Chakrabarty, the problem arose when bacteria grew in the same environment: "some strains tended to dominate over others, resulting in the survival of the fittest and, of course, overall loss of crude oil conversion."[3] But he understood that "plasmid-born" genes could be removed from one bacterium and transported into another, creating one multi-plasmid bacterium. Chakrabarty had an idea. What if he combined these bacteria to create one supreme oil-hungry bug?

During weekends and after his shifts at General Electric, Chakrabarty experimented with *Pseudomonas* genes to build a multi-plasmid bacterium. When he attempted to create a mixture of plasmids, he found that certain plasmids were incompatible with others. To get around the problem, he used ultraviolet irradiation to combine them into one large plasmid. He removed plasmids from three diverse types of bacteria and funnelled them into a fourth type, altering its characteristics. "I simply shuffled genes, changing bacteria that already existed," explained Chakrabarty. "It's like teaching your pet cat a few new tricks."[4]

General Electric understood the value of Chakrabarty's findings, and how easily they could be appropriated by its competitors. Anyone could scoop the engineered bacteria from the oil they were eating, take them back to a laboratory, and replicate them for their own purposes.

Appropriating other people's "property" is not a new phenomenon. In the mid 1700s, Richard Arkwright invented the water-powered spinning frame, which propelled England into the global market of cloth manufacturing. This was before patent legislation was introduced. To protect this invention, Parliament drafted specific regulations. The Arkwright machine and its workers were not to leave Britain. As of 1774, anyone who disobeyed was subject to a £200 fine and 12 years in jail. In 1790, an Arkwright worker named Samuel Slater, masquerading as a farmer, fled to the United States, where he re-created an Arkwright factory. This spurred America's own cloth manufacturing revolution, and Slater was named a hero—while, a continent away, he was considered a "patent infringer."[5]

In the U.S. in 1793, Thomas Jefferson introduced the Patent Act.[6] The original act defined a patentable invention as "any new and useful process, machine, manufacture, or composition of matter, or any new or useful improvement thereof. . . ."[7] Jefferson decreed that "ingenuity should receive a liberal encouragement."[8] As Justice Jackson of the U.S. Supreme Court stated in 1945, "The primary purpose of our patent system is not reward of the individual but the advancement of the arts and sciences. Its inducement is directed to disclosure of advances of knowledge which will be beneficial to society; it is not a certificate of merit, but an incentive to disclosure."[9] This incentive came in the form of cold, hard cash. Patents could be a scientific gold mine.

For an invention to be patentable in Canada, it must be new, useful, and non-obvious. A new invention is one that has never been "disclosed in a single source more than 12 months prior to the filing of the patent application." A useful invention needs to be of "industrial interest" and work properly. A non-obvious invention is a development or improvement that would not be obvious to someone skilled in the industry.[10] A patent filed before October 1989 protected an invention for 17 years. Inventions whose patents were filed after that year are protected for 20 years.

In Europe, patent holders must prove that their invention is not only new and inventive, but also morally justifiable. Article 6 of the EC Directive on the Legal Protection of Biotechnological Inventions[11] and Article 53 of the European Patent Convention[12] forbid the patenting of biotechnological inventions that are "contrary to *ordre public* or morality," which include cloning people and manipulating the genome of animals that "are likely to cause them suffering without any substantial medical benefit to man or animal." In October 2000, the European Patent Office stated that the morality restrictions prohibit the invention of "mixed-species embryos."[13]

While the European system focuses on an individual's basic right to own intellectual property, the United States views the patent system as a way to fuel industry.[14] Patented inventions can encourage investment in science, which can be costly and unpredictable. In biotechnology, it typically takes at least a decade to bring a product to market. In the meantime, billions are spent on R & D. The discovery stages of a drug are the most exorbitant. Companies must rely on deep-pocketed investors to fund their research, but investors won't bite if the competition has a good chance of getting to the pharmacy shelves first. So companies must ensure that no one else can use their ideas. Enter the patent—a "government grant of a time-limited legal monopoly given to an inventor in exchange for the public disclosure of an invention. It can be thought of as a veto over the activities of others in respect of making, using, selling or importing an invention."[15] It is not that a patent gives someone the right to produce, sell, or use a product; it prevents others from doing so. "If your invention is not protected," asks Chakrabarty, "why would [backers] invest money if somebody else can pick it up and copy it?"

In exchange for a 20-year patent, the inventor must divulge how the invention is made, so that someone "skilled in the art" can reproduce it. According to professor and author Jack Wilson, "The patent encourages disclosure of information that otherwise might not be open to scrutiny by other scientists or business competitors—in biotechnology these roles may be played by the same people."[16] Some scientists choose to forgo patents and rely on trade secrets to protect their inventions. But in the world of biotechnology, where it is always a distinct possibility that many other scientists are working on the same experiment, it pays to patent early and patent often.

---

There are two kinds of patents: a product claim and a process claim. Chakrabarty and General Electric filed for both: one for Chakrabarty's process of engineering bacteria, and one for the bacterium itself. The product claim quickly proved to be a problem, not to mention controversial. Bacteria are living matter, which most patent lawyers deem non-patentable under the U.S. Patent and Trademark Office's (PTO) regulations. But General Electric patent attorney Leo I. MaLossi did not concede. According to Chakrabarty, "MaLossi could not quite understand why an invention, which may otherwise qualify as a patentable invention, could not be patented simply because it was living." Chakrabarty further noted that in MaLossi's opinion, "a living microorganism is nothing but a composition of matter and genetic engineering techniques that have imparted a new and useful characteristic to this organism with a consequent change in the composition of its matter."

The PTO accepted the process claim but rejected the product claim, stating that microorganisms are non-patentable because they are products of nature. This had been established by previous case law. In 1948, the United States Supreme Court had decided in *Funk Brothers Seed Co.* v. *Kalo Inoculant Co.* that bacteria "are the work of nature. Those qualities are of course not patentable. For patents cannot issue for the discovery of the phenomena of nature."[17] The Funk Brothers Seed Company had been refused a patent on what it argued was an invention: the mixture of several types of bacteria. While the combination did not occur naturally, each separate bacterium did. In order for something to be considered an invention, the inventor must "add something to nature's handiwork."[18]

Chakrabarty had taken three types of naturally occurring bacteria and inserted them into a fourth. The act of insertion was a natural process that bacteria innately know how to perform. MaLossi argued that Chakrabarty's microorganism was not a natural product because the scientist had engineered the bacteria, creating something that could not be found in nature. The board of appeals accepted this, but maintained that the bacteria were alive and therefore non-patentable. General Electric decided to take its product claim to the U.S. Court of Custom and Patent Appeals (CCPA), which examines legislation covering patentable living plants.

The U.S. Plant Patent Act,[19] which was designed to protect the intellectual property of plant breeders, allowed for plants to be patentable inventions. If an individual discovered or invented a plant and could reproduce it asexually (using buds or cuttings rather than seeds to produce offspring), the living matter was patentable under this legislation. A patent on a plant forbids anyone but the patent holder to reproduce it using this method. When the plant patent legislation's merits were being debated, Luther Burbank, a famous plant breeder, explained its importance: "A man can patent a mousetrap or copyright a nasty song, but if he gives the world a new fruit that will add millions to the value of earth's annual harvests he will be fortunate if he is rewarded so much as having his name connected with the result."[20] American plant protection legislation aims to "remove existing discrimination between plant developers and industrial inventors" rather than separate naturally occurring products from manufactured ones.[21]

But are living organisms patentable like asexually reproducible plants? Referring to Chakrabarty's patent application, Chief Judge Howard Markey of the CCPA stated that "The sole issue before us is whether a manmade invention, admittedly novel, useful, and un-obvious, is non-patentable because and only because, it is 'alive' (in the sense that microorganisms are 'alive')."[22] In 1978, Justice Giles Rich held that microorganisms are in fact patentable: "The fact that microorganisms . . . are alive . . . [is] without legal significance." The CCPA found that microorganisms are patentable because they are "more akin to inanimate chemical compositions such as reactants, reagents, and catalysts, than they are to horses and honeybees or raspberries and roses."[23] Unsatisfied with the decision of the CCPA, the PTO appealed to the U.S. Supreme Court, which reviewed the case. In a previous decision, *Parker* v. *Flook,* the Supreme Court had noted, "we must proceed cautiously when we are asked to extend patent rights into areas wholly unforeseen by Congress."[24] The CCPA ignored the Supreme Court's warning, and reaffirmed its decision that microorganisms are patentable. Still unsatisfied with the CCPA's decision, the PTO again asked the U.S. Supreme Court to re-evaluate the case.

By this time, General Electric had set aside plans to manufacture the bacteria. Laboratory tests had concluded that while the bacteria could break down small amounts of oil, even this multi-plasmid variety was incapable of stomaching a vast oil slick. But this didn't stop General Electric from revisiting the case. There was much more at stake than a single engineered microorganism. If the Supreme Court favoured Chakrabarty, an entirely new patent industry would be born. Researchers and investors alike predicted that the patenting of cells, genes, and proteins would create a genomic gold rush. Companies would be able to take more chances on biotech when they could offer investors the security of patent protection. It was a chance to profit on life itself.

In 1980, the U.S. Supreme Court heard oral arguments in the case of *Diamond, Commissioner of Patents and Trademarks v Chakrabarty*.[25] "The briefs present a gruesome parade of horribles," the court learned. "Scientists, among them Nobel laureates, are quoted suggesting that genetic research may pose a serious threat to the human race, or, at the very least, that the dangers are far too substantial to permit such research to proceed apace at this time. . . . [A]t times, human ingenuity seems unable to control fully the forces it creates—that, with Hamlet, it is sometimes better 'to bear those ills we have than fly to others that we know not of.' "[26] But this reasoning was rejected by the U.S. Supreme Court, which explained that such moral and political ambiguities should be addressed not by the courts, but rather by Congress. On June 16, 1980, after eight years of deliberations, the Supreme Court made its ruling in the Chakrabarty case.

Chief Justice Warren Burger, writing for the court, explained that the question to be answered was a narrow one. Rather than attempting to resolve the moral issues of the case, the court's only task was to consider whether Chakrabarty's product claim should be accepted. Chief Justice Burger explained that the task of the court involved an interpretation of Section 101 of the United States Code (U.S.C.), which states, "Whoever invents or discovers any new and useful process, machine, manufacture, or composition of matter, or any new and useful improvement thereof, may obtain a patent therefor, subject to the conditions and requirements of this title."[27] In other words, the court's role was to determine whether Chakrabarty's bacteria constituted a "manufacture" or "composition of matter." The product's novelty or non-obviousness was not being questioned.

The court looked to the dictionary definition of "manufacture," finding that it meant "the production of articles for use from raw or prepared materials by giving to these materials new forms, qualities, properties, or combinations, whether by hand-labour or machinery."[28] "Composition of matter," stated Chief Justice Burger, was defined as "all compositions of two or more substances and . . . all composite articles, whether they be the results of chemical union, or of mechanical mixture, or whether they be gases, fluids, powders or solids."[29] The court concluded that, in choosing such "expansive" terms as "manufacture" and "composition of matter," Congress intended to give patent laws a wide scope.

But did this wide scope include products of nature? That was the vexing question that had plagued the courts for nearly a decade. The reason, says Chakrabarty, had to do with God. "The concept was that anything that is natural cannot be patented. It needs human intervention. People at that time thought that all living creatures are creations of God," he explains. "So they're natural. Until they learned how to manipulate living forms and then they were not exactly the products of nature." If, then, human intervention could genetically alter something that was living, should it be deemed patentable? Chief Justice Burger reflected on the rewritten patent laws of 1952, when Congress replaced "art" with "process." He held that this change in language was evidence of an intention on the part of Congress that patentable inventions should "include anything under the sun that is made by man."[30]

In a five-to-four decision, the U.S. Supreme Court decided that Chakrabarty's microorganism was in fact not a product of nature, but rather a product of his own hands. Chakrabarty had altered the naturally occurring bacteria to engineer his own microorganism. Chief Justice Burger explained, "Thus, a new mineral discovered in the earth or a new plant found in the wild is not patentable subject matter. Likewise, Einstein could not patent his celebrated law that $E=mc^2$; nor could Newton have patented the law of gravity. Such discoveries are 'manifestations of . . . nature, free to all men and reserved exclusively to none.'"[31] On the other hand, held the court, "Here, by contrast, the patentee has produced a new bacterium with markedly different characteristics from any found in nature and one having the potential for significant utility. His discovery is not nature's handiwork, but his own. . . ."[32]

Chakrabarty believes that the Supreme Court decided in his favour because it kept in mind that a fundamental intention of the U.S. Constitution is to encourage innovation. After the Chakrabarty decision, the PTO was flooded with life patent applications. "Scientists began to isolate, snip, insert, recombine, rearrange, edit, programme, and produce biological genetic material," Chakrabarty said in 2002. "In fact, scientists for the first time began to have the potential to become the architects of life. . . . Body parts and genetic materials began to be sold and patented, manipulated and engineered. An unprecedented change was observed in many of our most basic social and legal definitions. Scientific and legal jargons obscured important moral distinctions."[33] In 1980, Genetech's IPO had the fastest price-per-share rise that Wall Street had ever seen, shooting up from US$35 to US$89 in 20 minutes, generating "paper fortunes" of at least US$50 million. Genetech proclaimed, "The court has assured the country's technology future."[34] Cetus set its own record, generating the most cash of an IPO in history: US$115 million. Two American patent lawyers pointed out that the Chakrabarty decision fell on the centennial anniversary of the patent approval for Edison's incandescent lamp. General Electric's win, they felt, would have greater influence on "the living standards in the 21st century . . . than the granting of the light bulb to Edison in 1880 had on the 20th century."[35]

Even though Chakrabarty's was a lowly microorganism, it seemed completely plausible that other living organisms would soon be patented. Just as the printing press or personal computer could be monopolized, so could life. "The legal distinctions between life and machine, between life and commodity, are gradually beginning to vanish," says Chakrabarty.[36] People began posing the vexing question, What is life? But now the question had become more complicated. Is it something sacrosanct and beyond value? Or is it merely a product to be manipulated, manufactured, and profited from? Could individuals invent new life and patent it as their own? And if they could, where would we go from here?

---

In 1987, the lifespan of patents was again extended. The U.S. Patents and Trademark Office overrode earlier regulations on the patenting of living organisms, and declared that all "genetically engineered multicellular living organisms" were patentable. For organisms to be deemed patentable, said Donald J. Quigg, Assistant Secretary of Commerce and Commissioner of Patents and Trademarks, "they must be given a new form, quality, properties, or combinations not present in the original article existing in nature."[37] The PTO was quick to point out that this did not extend to humans. The 13th Amendment to the Constitution prohibits human slavery, or property rights over individuals. Yet the parts of a human—cells, genes, proteins, and tissue—were patentable, as were animals. In 1988, the world got a glimpse of the first patented mammal.

The OncoMouse, more famously known as the Harvard mouse, was invented by Harvard researchers Philip Leder and Timothy Stewart. These scientists shuffled the animal's genetic makeup by inserting "myc," a cancer-causing gene, into its embryo, creating an engineered mouse predisposed to cancer. When the OncoMouse was mated with a normal mouse, half the offspring carried the cancer-causing gene. The OncoMouse was designed as a test model for research. The carcinogenic properties of materials could be tested in the mice in small doses proportionate to the amount humans come into contact with, and the mice would develop tumours at an accelerated rate, speeding up analysis.

But the OncoMouse raised the hackles of animal activists around the world. Were individuals to be given the right to engineer a life of pain, one that would surely end with a losing battle against cancer? Had life been stripped of its sanctity and left with nothing more than utility? U.S. Senator Mark Hatfield asked, "Will future generations follow the ethic of this patent policy and view life as mere chemical manufacture and invention with no greater value or meaning than industrial products? Or will a reverence for life ethic prevail over the temptation to turn God-created life into reduced objects of commerce?"[38]

Not only did Harvard and its financial backer, E.I. DuPont, want to patent the gene insertion method, they also sought a patent on any non-human mammal and its descendants engineered by this gene manipulation. On April 12, 1988, when the PTO issued Patent No. 4736866, DuPont received an astoundingly broad monopoly. Not only was the company granted U.S. rights over the furry little rodent; it would also control any species "of transgenic nonhuman mammal all of whose germ cells and somatic cells contain a recombinant activate oncogene sequence introduced into said mammal, or an ancestor of said mammal, at an embryonic stage."[39] Not only could it engineer mice predisposed to cancer, DuPont had the sole rights to manufacture cats, goats, apes—whatever non-human mammal it chose—with that predisposition.

Harvard University's Ruth Hubbard and Tufts University professor Sheldon Krimsky explain that the Harvard mouse patent is not for one specific mouse, but rather for a strain of mice, in which no two animals are identical. "The patent covers the genetically modified mouse, all of its progeny and all of their progeny for 17 years," they state, "much longer than the life span of the 'manufactured' mouse. Can Drs. Leder and Stewart really be said to have manufactured generations of mice because they operated on the eggs of one?"

After securing its U.S. patent, Harvard won patent approval in 17 other countries, including France, Germany, Greece, Ireland, Italy, and the United Kingdom. But Canada was not so easy to convince. On June 21, 1985, the Canadian patent was filed. On February 21, 1990, the patent commissioner rejected 18 of Harvard's 24 claims, with 17 of these being outside the scope of the definition of a patentable invention under the Canadian Patent Act.[40] The process claim was granted because the act of inserting the myc gene made the mouse useful for cancer research, fulfilling the utility requirement. Harvard was thus given a Canadian patent covering the engineering of the plasmid used, as well as the method used to transplant it into the animal. But the mouse itself proved tricky for Canadian courts to categorize.

The definition of "invention" under Canada's Patent Act is almost identical to Thomas Jefferson's. Section 2 provides that a patentable invention is "any new and useful art, process, machine, manufacture or composition of matter, or any new and useful improvement in any art, process, machine, manufacture or composition of matter." The question vexing Canadian courts in the case of the Harvard mouse focused on the difference between lower and higher life forms. The Canadian Patent Office defines higher life forms as "multi-cellular differentiated organisms (plants, seeds, animals)" and regards them as non-patentable.[41] Harvard appealed to the Federal Court, Trial Division, which agreed with the Commissioner of Patents. Harvard then appealed to the Federal

Court of Appeal, which approved the appeal and granted the controversial patent without coming to a conclusion regarding the distinction between higher and lower life forms. This compelled the Commissioner of Patents to appeal the decision of the Federal Court of Appeal. On June 14, 2001, the case was sent to the Supreme Court of Canada to end the debate.

In Canada, the question of patenting higher life forms turns on whether the definition of "invention" under the Patent Act is sufficiently broad to include higher life forms. Are life forms such as the OncoMouse either a "manufacture" or a "composition of matter"? If the Supreme Court of Canada decided that they were, warned Justice Michel Bastarache, speaking for the majority of the court, they would be creating a slippery slope towards patenting all forms of life, including human life: "Should this Court determine that higher life forms are within the scope of Section 2, this must necessarily include human beings. There is no defensible basis within the definition of invention itself to conclude that a chimpanzee is a 'composition of matter' while a human being is not."[42] Justice Bastarache made a distinction between higher and lower life forms, stating that there was a clear difference between bacteria and fungus, and animals: "The distinction between lower and higher life forms, though not explicit in the Act, is nonetheless defensible on the basis of common sense difference between the two. Perhaps more importantly, there appears to be a consensus that human life is not patentable; yet this distinction is also not explicit in the Act. If the line between lower and higher life forms is indefensible and arbitrary, so too is the line between human beings and other higher life forms. . . ."[43] Justice Bastarache found that the definition of "invention" did not include higher life forms, concluding that if the mouse patent was granted, it would be difficult in the future to decide what life forms were and were not patentable.

The Federal Court of Appeal had simplified the issue by not deciding what constituted higher life forms. How was a human to be defined, and did a ban on patenting human beings extend to human fetuses and organs? Justice Bastarache noted, "A complex life form such as a mouse or a chimpanzee cannot easily be characterized as 'something made by the hands of man.' Nor is OncoMouse a 'composition of matter.' Higher life forms are generally regarded as possessing qualities and characteristics that transcend the particular genetic matter of which they are composed. A person whose genetic makeup is modified by radiation does not cease to be him or herself. Likewise, the same mouse would exist absent the injection of the oncogene into the fertilized egg cell; it simply would not be predisposed to cancer."[44]

Justice Ian Binnie, speaking for the minority, dismissed the danger of patenting human beings, stating that this was forbidden under the Canadian Charter of Rights and Freedoms.[45] Justice Binnie went on to state that it was "ludicrous" for Canada to stand alone and refuse the OncoMouse its patent.

Nevertheless, that's just what Canada did. On December 5, 2002, the five-to-four decision made headlines around the world. Canadian courts would stand their ground in the world of patenting life forms; after 17 years of debate, the OncoMouse was judged non-patentable in Canada. So were the 1,500 applications that had been on hold for two years pending the Supreme Court's decision: claims including a genetically altered salmon that could grow at an accelerated rate, and a goat that could make milk using proteins manufactured by drug companies.[46]

The scientific and moral complexities of the day continue to challenge patent legislation. Patents were initially conceived as a legal way to stimulate learning and scientific progress that would result in commercial and medical benefit for all. When the world's first patent acts were being drafted, no one conceived of mice genetically engineered to develop cancer. Those responsible for drafting such legislation probably never imagined the complex reality of companies wielding control over science, and controlling global health. One such company is Myriad Genetics.

---

For over 15 years, a scientist named Mary-Claire King pursued the elusive breast cancer gene she had named BRCA1. In 1974, King joined a laboratory at the University of California, San Francisco, and dedicated herself to studying the disease, which, according to the Canadian Cancer

Society, afflicts one in nine Canadian women.[47] King's approach was to gather data on families with a history of early-onset breast cancer as well as ovarian cancer. If there was a hereditary element to the disease, these families would reveal it. Her critics felt this was an unjustified assumption, because breast cancer is caused by a mixture of genes and lifestyle factors. Yet by 1988, King believed that her 1,579 at-risk families were leading her to a genetic bull's-eye.

In 1990, King's persistence paid off. Combing through the long grass of DNA, she cornered BRCA1 on chromosome 17; the elusive gene was hiding in a dense thicket of DNA along the long arm of the chromosome. She had unearthed a restriction fragment length polymorphism (RFLP), or genetic marker, that was connected to breast cancer in 23 of her families, with 146 cases of the disease over three generations (for an explanation of RFLPs, see Chapter Two). King ensured that the female family members had not been exposed to an excess number of X-rays.[48] Her data were solid, and King was exhilarated. Maybe these findings would help the estimated 21,600 women diagnosed annually in Canada. Of these women, 5,300 will lose their lives to breast cancer, the second leading cause of death in women, behind heart disease.[49] If women know they are at risk of developing the disease, they can be monitored closely and have the option of early and aggressive treatment before it is too late.

The BRCA1 gene gives instructions to make a protein that diligently repairs DNA. If the BRCA1 gene is switched off, this protein becomes jumbled, disrupting normal cellular function. Mutations of the breast cancer gene happen when there are large deletions or flips within the string of DNA. The body tries to read the altered genetic sequence, which results in a malfunctioning protein. BRCA1 can be thought of as a tumour suppressor gene. When its genetic sequence becomes shuffled and its proteins are non-functioning, cancerous cells may result. There have to be two mutations of the BRCA1 gene in order for cancer to appear. But someone who inherits a mutation from only one parent, and still has one functioning copy of the gene in every cell of her body, is still predisposed to breast as well as ovarian cancer—because a random mutation can knock out the remaining functioning copy of the gene, initiating the disease. In that case, the gene's tumour suppression is silenced, and cancer cells begin to multiply uncontrollably. While the BRCA genetic mutations are not the only cause of breast and ovarian cancers, those who carry them have a 40 percent to 85 percent chance of developing breast cancer, and a 16 percent to 40 percent chance of developing ovarian cancer. The offspring of carriers have a 50 percent chance of inheriting the mutation.[50] But carriers of only one mutation may never develop cancer; as long as the second, functioning copy of the gene remains intact, they may remain free of the disease.

King proudly announced at the 1990 American Society of Human Genetics Meeting that she had narrowed the location of the BRCA1 gene to chromosome 17. She also shared her findings with the Breast Cancer Linkage Consortium (BCLC), a database that enables public and private organizations to analyze the work of scientists like King.

But King and her team, which now included renowned gene hunter Francis Collins, were not the only ones doggedly pursuing the exact location of the BRCA1 gene. A Salt Lake City biotech firm had quietly entered the genetic scavenger hunt, and was closing in on first prize. Mark Skolnick, a population geneticist from Utah, and Wally Gilbert, co-founder of biotech company Biogen, had had an idea. They would form a gene-mapping company that would use the lineages of Mormon families, which originate from a relatively small number of ancestors. It would be called Myriad Genetics. In no time, the pair was plotting a course to nab the breast cancer gene. In August 1994, Skolnick and his group, along with teams from the University of Utah, the NIH, and McGill University in Montreal, ended the hunt. Using the data from the BCLC as a jumping-off point, they were able to sequence the BRCA1 gene. The project was funded by private money from Eli Lilly as well as public organizations, including more than US$5 million from the NIH. Myriad filed for U.S. "composition-of-matter" and "methods-of-use" patents on the BRCA1 gene as well as on the many mutations it had uncovered.

Although her own project had been suddenly deflated and two decades of her work appropriated, King remained optimistic that the mapping of the breast cancer gene could save the lives of countless women. At a news conference following Myriad's announcement, she said, "In the 20 years since we have been working on this project, more than a million women have died of breast cancer. We very much hope that something we do in the next 20 years will preclude another million women dying of the disease."[51] But Myriad had other plans for its patent. The company focused on developing a test to screen women for BRCA mutations. Overnight, Myriad had a priceless monopoly on breast cancer screening. The first year it exercised its patent, it had a US$1.5-million increase in revenues from genetic testing.[52]

Then it was discovered that BRCA1 had a partner in crime: BRCA2, its sister gene. In September 1994, the second breast cancer gene had been roughly located on chromosome 13, but was yet to be isolated. Myriad was back in the race, competing against scientists around the world. The final stretch had the biotech firm neck-and-neck with a group of U.K. researchers. Michael Stratton and teams at the Institute for Cancer Research and the Sanger Centre thought the BRCA2 patent was theirs. On December 22, 1995, the U.K. group invited the press to announce the BRCA2 sequence as well as the filing of their U.K. patent. That same day, Myriad blindsided the British researchers by announcing that it had already discovered BRCA2. The U.K. group was too late. The U.S. patent had already been filed, and the Myriad monopoly secured.

Prior to 1996, screening was available only on a research basis and at no cost. [53] But in 1996, Myriad began marketing its BRCA1 test, at US$900. Not long after, the test was recalled because of resounding condemnation over the lack of genetic counselling, which could result in misinterpretation of test results. So the company repackaged BRACAnalysis as a laboratory test to be administered by a physician, who would explain the results, lessening the chance of misunderstandings. Myriad also compiled educational materials on its testing methods and the realities of breast and ovarian cancer, which are accessible via the Internet.

Myriad now offers several breast cancer tests. The single-site BRACAnalysis screens for a single mutation in a family and costs CDN$525. Its multi-site BRACAnalysis rings in at CDN$600 and tests for three mutations common among Ashkenazi Jews. Myriad's comprehensive

BRACAnalysis, which does full gene sequencing, costs CDN$3,850, and is required by those whose families have never had any BRCA screening. The comprehensive test sequences both BRCA1 and BRCA2 genes to search for mutations. Once a family mutation is recognized, other family members are given the choice of undergoing the CDN$525 single-mutation test to determine whether that particular genetic glitch has been passed along.

Myriad's tests use very sophisticated DNA sequencers, which may account for the high cost of screening. The company insists that its methods are the "gold standard" of genetic testing.[54] Base pairs as fine as blades of grass are exhaustively checked along the BRCA genes, revealing mutations hiding in the thick genetic undergrowth. The Institut Curie, France's eminent cancer research centre, has challenged this. According to Dr. Dominique Stoppa-Lyonnet, who has long been associated with the Institut Curie, Myriad has a very good method for the detection of small mutations, but does not search for large but partial deletions or duplications of the BRCA1 gene. Thus it runs the risk of missing about 15 percent of the mutations.[55] A paper published by the Institut Curie explained, "In order to get the most reliable results, geneticists cannot simply go for results derived from direct sequencing; they need to be able to carry out additional research using more comprehensive technologies. Through its monopoly position Myriad, which only uses direct sequencing technology for initial searches of family mutations, would seriously jeopardize the quality of test results, as 10 to 20 percent of all mutations would thus go undetected."[56]

But the reality is that Myriad's patents secure its rights to the BRCA1 and BRCA2 genes. If scientists want to experiment within a section of patented DNA, they must obtain permission from the patent holder. Contracts and permission fees must be exchanged before researchers can get their hands on such DNA. If these scientists then make a discovery along such DNA—say they find a new protein predisposing the person to a certain disease—they must make royalty payments for the derivation of the new protein to those who own the rights to the genetic sequence.

In Canada, BRCA mutation screening was made available in 1996 on a research trial basis. In British Columbia clinical testing was offered that year by the Hereditary Cancer Program (HCP). Other publicly funded programs were launched across the country to test for mutations, including those predominant in French-Canadian and Ashkenazi Jewish populations. But the public system in Canada has been unable to meet the demand for screening.

In 1999, outraged by the dangerously long waiting lists, an Ontario woman informed the Ontario Health Insurance Plan (OHIP) that BRCA testing was an essential medical service, and that inadequate access was a danger to her health. Even though Fiona Webster was at risk of inheriting breast cancer, she was denied testing because she did not meet testing criteria and she was not enrolled in a research study. OHIP farmed out Webster's testing to Myriad so that she could be screened for breast cancer quickly, to decide if she should have preventative surgery. In March 2000, the Ontario government announced that it would fund BRCA screening for hereditary breast and ovarian cancer. But in October of that year everything changed. Myriad was given its Canadian patents on the BRCA1 gene. In April 2001, it was awarded its patents on the BRCA2 gene. Public institutions that had been offering testing free of charge to patients were sent "cease and desist" letters informing them that all in-house screening was to end. From this point on, Myriad would be performing all BRCA analysis. Public agencies that funded initiatives like British Columbia's HCP did not have budgets big enough to pay for Myriad's tests. HCP was doing screening for $1,200 per patient, compared to the $3,850 Myriad was suddenly charging. HCP patients were told they had to foot the bill if they wanted to continue with the testing. Then, in February 2003, the B.C. Ministry of Health Services decided to ignore Myriad's patents, and told the HCP and the B.C. Cancer Agency to proceed with their in-house screening.[57]

The Ontario government refuses to pay Myriad's fees, stating that its Canadian patent does not warrant such monopolistic demands, and plans to settle the dispute in court. In the meantime, Ontario is performing its own BRCA screening, which costs the government $1,100 per test.[58] A similar scenario ensued in the U.S. Once Myriad had secured its BRCA1 patent, it informed staff at the University of Pennsylvania that they must pay every time they used the test. Pennsylvania's Genetic Diagnostic Laboratory had been using Myriad's tests for a study backed by the National Cancer Institute. Arupa Ganguly, director of research and development at the university's Genetic Diagnostic Laboratory, has said, "In reality, this lab has been stopped from testing for BRCA1 and BRCA2."[59]

Over the years, Myriad has flexed its patenting muscles around the world. It now has global dominance over breast cancer screening, with patents in the U.S., Canada, the European Union, New Zealand, and Australia. The company forbids academic researchers to use its patented genetic sequences if they are going to develop other screening methods for the disease. As well, data from participants in academic studies are never revealed to the participants, because such diagnostic clinical use of the Myriad tests would be a violation of Myriad's patents.[60]

Members of the medical community are voicing their outrage at the company's monopoly. At its 2002 annual general meeting, the Canadian College of Medical Geneticists voiced its concern that gene patents do not recognize the collaborative work and public investment that go into patented discoveries. These patents may also restrict benefits that could arise from genomic research. The college stated that "unreasonable exploitation of the entitlements of a patent holder will be detrimental to the health and well-being of Canadians." It warned that life patents must not create commercial monopolies or genetic databanks with restricted access, and that these patents must not restrict the public health care system's access to the latest medical advances.[61] The American College of Medical Geneticists has gone further, asking for a ban on the patenting of genes, on the basis that it results in astronomical user fees and "monopolistic" licensing.[62]

The Institut Curie has been extremely vocal in its criticism of Myriad's dominance. In a 2002 paper, it analyzed the scope of the company's patents. In Europe, Myriad's patent EP 699 754 encompasses any methods that diagnose a predisposition to breast and/or ovarian cancer using the normal sequence of the BRCA1 gene. Its patent EP 705 903 concerns specific mutations and the diagnostic methods to detect such mutations. Its patent EP 705 902 covers diagnostic kits. Any scientist who infringes on these patents, explains the Institut Curie, is subject to prosecution, regardless of which "detection technology" is being employed. Even if they are not using Myriad's direct sequencing method, laboratories are forbidden to perform any type of predisposition screening for BRCA1 and BRCA2 mutations, and must send their DNA samples to Myriad's "testing plant" in Salt Lake City. The danger, says the institute, is that this will enable Myriad to compile the only genetic databank on breast and ovarian cancer in the world. "This monopoly on patent exploitation will lead to a loss of expertise and information among physicians and research scientists in Europe, as they will no longer be allowed to improve diagnostic technologies and methods," warns the Institut Curie.[63]

Over the years, scientists around the world have refused to abide by Myriad's rules or to respect its patent claims. Some European laboratories have chosen to continue using their own testing methods, which have been in development for years and can be performed at a fraction of Myriad's price. In 2001 the Institut Curie's Stoppa-Lyonnet was quoted as saying, "We refuse to send [Myriad] DNA. I am against monopolies on genetic testing. It's not good for patients. And it's not good for the field."[64] Critics feel that Myriad's patents have slowed research on breast and ovarian cancer. And what about the countless women who cannot afford these fees for disease screening? This may cost them their lives. Until Myriad's patents are lifted, scientific progress on these deadly diseases has effectively been frozen. "The substance patents now being given to the human genome are inappropriate and endanger research and medicine," says Dr. Otmar Kloiber, secretary-general of the World Medical Association. "Information about the human genome can't be invented. It is the common heritage of all humans."[65] Yet this common heritage is being appropriated.

---

A metaphor is sometimes used to explain patent control over future research. By patenting the BRCA1 and BRCA2 genes, Myriad controls knowledge close to its source. Think of a stream of water. If a dam is built where the water begins to flow, no one downstream can share it. Filing a patent can be similar to building a dam; if the company holding the patent refuses to freely share its precious resource, downstream products like new vaccines and treatments remain undiscovered and undeveloped. And any errors within the dammed-up knowledge go undetected, without other scientists examining and verifying results.[66]

Almost 1,000 life patents have been awarded to Incyte Genomics, Celera Genomics, Hyseq, Millennium, and Human Genome Sciences, which collectively have applied for another 25,000. As author Matthew Albright points out, "Considering that the human being has only about 32,000 genes, that's a lot of upstream territory being grabbed up."[67] While law professor Timothy Caulfield, who specializes in health law, agrees that gene patents pose a danger of stifling research because they promote secrecy among scientists who are vying for the same patents, he feels they are a necessary evil. "They're obviously incredibly important to the biotech industry that is so fragile," he says. "The payoffs are maybe 10, 12 years away. You need a huge amount of upfront capital to make this work, and a patent is one of the only forms of property that gives them a sense of security."[68]

However, should there be such a thing as property rights over human DNA, the collective inheritance among all of humanity? Is it truly an invention if someone discovers what has, after all, always been hiding inside each and every one of us? It has been argued that the human genome should not be patentable because of its special status. It is unique and distinctive, and should be treated with greater respect than what is accorded to the genomes of animals or plants.[69] According to London's Nuffield Council on Bioethics, patenting a particular gene for a specific use constitutes claiming a discovery of an "association"—the uncovering of a pre-existing fact—rather than an invention, which is the creation of something novel.[70] Should property rights be issued for the pre-existing tissue, cells, genes, and proteins that make up every individual? It has long been held that an entire body is non-patentable because the patent would infringe on human dignity, which, according to the United Nations

Declaration on Human Rights,[71] is the foundation of human rights. For human dignity to be upheld, individuals should not be viewed as commodities. According to the Canadian Biotechnology Advisory Committee, "Even if the act of granting a patent on an invented human were not in itself a violation of basic human rights, exercising the patent's exclusive right to make, use, or sell an invented human would almost certainly violate the *Canadian Charter of Rights and Freedoms* and the *Canadian Human Rights Act*."[72]

---

In 1976 a gravely ill seafood salesman named John Moore met with doctors at the UCLA Medical Center who had discovered that Moore was suffering from hairy cell leukemia, a rare form of cancer that is often fatal. Moore was informed that his life was at risk. The cancer had attacked his spleen, which had ballooned from 500 grams to almost six kilograms. Dr. David W. Golde, the attending physician, advised that Moore's spleen be removed.

While Moore consented to the splenectomy and gave the university rights to his organ, he was unaware that his doctor had a special interest in his tissue. Golde and UCLA researcher Shirley Quan suspected that Moore's spleen cells had the ability to produce a unique blood protein that could be used to create an anti-cancer agent. The spleen was removed and a section of the organ was transported to a UCLA laboratory, where it became the focus of intensive research. The UCLA team cultured Moore's cells and produced a cell line from his white blood cells, aptly naming it Mo. The line produced T-cell growth factor interleukin-2 and immune interferon 2, which are useful in the fight against cancer. After his operation, Moore returned to UCLA for several checkups. Golde insisted that his patient be examined at his facility, under his supervision. When Moore was unable to pay the airfare to get back to Los Angeles, Golde paid for his tickets. During each checkup, Moore gave samples of blood, skin, bone marrow, and sperm. In March 1984, unbeknownst to Moore, Golde and Quan were approved for a patent on Moore. Overnight, he became patent #4,438,032.

The patent afforded Golde many financial opportunities. Its potential for generating cash had secured him contracts even in the years leading up to its approval. He had signed with Genetics Institute, Inc., and Sandoz Pharmaceutical Corp. and its related entities, drafting plans to profit from the Mo patent. Golde was handed 75,000 shares of Genetics Institute at a "nominal price," while Sandoz and Genetics Institute paid Golde and UCLA $440,000 over three years. At one point Golde's stock was worth $3,000,000, and potential products that could be derived from the Mo cell line had an estimated value that was many times this amount. [73]

When Moore learned of the patent, he was shocked and outraged. Without his knowledge, he had been used as a lab experiment. "I thought he was my friend," he said of Golde. "He talked about bringing his kids up to go fishing with me near my place in Seattle. I said I'd like that."[74] He thought back to all those post-operative checkups, and decided that his medical team had not acted solely in the interests of his health; they had also been motivated by the almighty dollar. Moore filed a suit in the Superior Court, Los Angeles County, alleging that the UCLA medical team had misappropriated his "bodily substances" and breached his ownership rights over his own tissue. He charged Golde, Quan, the university, and several biotech firms with fraud, deceit, unjust enrichment, the illegal appropriation of his property, and failing to obtain informed consent.[75]

Even though Moore had handed the rights to his spleen over to the university when it was removed from his body, and had consented to many vials of blood being drawn after the operation, he had done so without the knowledge that his organ would be used commercially. Moore's legal team alleged that he had never given true informed consent, since Golde and Quan had never revealed their plans. Had Moore been informed that research using his own cells might one day generate a lucrative product, he would have asked for remuneration. Allen B. Wagner, who acted as an attorney for the university, stated that the purpose of informed consent was to ensure that a patient understood the risks and choices available in relation to a medical procedure. Wagner explained that the intention behind informed consent was to "promote two values—one is patient well-being and the other is patient self-determination," not to confer a profit.[76]

Moore, however, maintained that he had been deceived. Sanford Gage, one of the attorneys representing him, argued, "The central issue in this case is the patient's right to a share of the profits earned by drug companies and biogenetic firms from products derived from his body."[77] The university thought otherwise. According to court documents, it argued that "Moore has no right, title or interest in the patent, notwithstanding the fact that his diseased spleen tissue presented a circumstance that stimulated the intellectual curiosity and ingenuity" of the UCLA researchers. The university pointed to another patent case in which a court had stated, "patents are not granted for the natural properties inherent in things existing in nature, although they may be granted for things an inventor does with those properties."[78]

Golde maintained that he had informed Moore of "everything he needed to know," which did not include his plans to patent the patient's cell line. The court agreed, stating that an individual's property rights do not extend to tissue that has been used. Golde said he was "surprised" at Moore's lawsuit. "This patient came to an expert institution. We helped him and we developed some important science. But the science was not something that was in his body. A patent is given for intangibles, not a thing. The idea was the use of a cell line."[79] But in July 1988, the California Court of Appeal reversed the decision of the trial judge, stating that Moore did in fact exercise part ownership to the patented cell line developed from his spleen. The medical and scientific communities were in shock. What would this mean for all the existing products that had been created from people's tissue? Patent holders were dismayed by the vision of a never-ending queue of patients demanding back payment for years of profit derived from scientific experiments performed on their tissues.

In the California Court of Appeal decision, writing for the majority, Justice Rothman held that the "Defendant's position that the plaintiff cannot own his own tissue, but that they [the University and the biotech companies] can, is fraught with irony. Apparently defendants see nothing abnormal in their exclusive control of plaintiff's excised spleen, nor in their patenting of a living organism derived therefrom. We cannot reconcile the defendants' assertion of what appears to be their property interest in removed tissue and the resulting cell-line with their contention that the source of the material has no rights therein."[80] Yet Justice George warned that this could lead to the uncontrolled marketing of human parts. He stated, "The absence of legislation regulating the trafficking in human body parts raises the spectre of thriving 'used body parts' establishments emulating their automotive counterparts, but not subject to regulation comparable to that governing the latter trade."[81]

The debate was taken to the Supreme Court of California. In *Moore* v. *Regents of the University of California,*[82] the Court decided that while Moore should have been informed of UCLA's intentions regarding his spleen cells, he had no right to profit from their patented discovery. Once Moore's spleen had been removed, he no longer exercised property rights over it. The Supreme Court stated:

> Research on human cells plays a critical role in medical research. This is so because researchers are increasingly able to isolate naturally occurring, medically useful, biological substances and to produce useful quantities of such substances through genetic engineering. These efforts are beginning to bear fruit. Products developed through biotechnology that have already been approved for marketing in this country include treatments and tests for leukemia, cancer, diabetes, dwarfism, hepatitis-B, kidney transplant rejection, emphysema, osteoporosis, ulcers, anemia, infertility, and gynaecological tumours, to name but a few.
>
> The extension of conversion law into this area will hinder research by restricting access to the necessary materials. Thousands of human cell lines already exist in tissue repositories, such as the American Type Culture Collection and those operated by the National Institutes of Health and the American Cancer Society. These repositories respond to tens of thousands of requests for samples annually. Since the patent office requires the holders of patents on cell lines to make samples available to anyone, many patent holders place their cell lines in repositories to avoid the administrative burden of responding to requests. At present, human cell lines are routinely

copied and distributed to other researchers for experimental purposes, usually free of charge. This exchange of scientific materials, which still is relatively free and efficient, will surely be compromised if each cell sample becomes the potential subject matter of a lawsuit.[83]

The Moore case has been considered in several subsequent U.S. decisions. Moore himself spoke out about it. "How can it be right that I trusted someone to help me and then I find out that they have secretly profited from part of my body?" he asked. "Why can someone else profit from my cells but I can't? Who really owns my body?"[84] This is a complex and difficult question to answer. Should patients receive profit from their own tissue, simply because they once housed it? People who have valuable body parts, such as naturally occurring rare protein types, would probably never learn of their genetic wealth unless a physician revealed their potential biological windfall. And if this physician was the one who discovered the utility of that body part, should its owner be compensated for producing it? Should people be encouraged to sell body parts at all?

Lawyer and commentator George Annas noted that it all came down to money. He said of the Supreme Court of California's decision, "California's courts decided who could reap profits from a cell line . . . as is clear from the text, the majority simply accepts the Chicken Little argument that if John Moore's property interest in his cells is upheld, the biotechnology industry's sky will fall on them and medical progress will suffer a major setback. In this regard the justices seem to have been blinded by science and unable or unwilling to distinguish it from commerce. The court essentially concluded that the biotechnology industry is both wonderful and fragile. Since it is wonderful, we must do our part to foster it; since it is fragile, we must protect it from harm. . . . In the court's flowery words, recognizing [Moore's property claim] would threaten to destroy the economic incentive to conduct important medical research."[85]

The Moore case was emotionally charged because it dealt with concepts of personal property, which are inextricably linked to concepts of self. Our bodies, which constitute our own personal property, can also be viewed as our individual selves. Fungible property, on the other hand, is something found outside the self. While it is generally acceptable for people to profit from their own fungible material, it is undecided whether making money from the property of the self is morally acceptable.[86] The Moore case showed that it is legally acceptable for people to patent and profit from other people's cells, yet the ethical debate continues. Was Moore himself degraded by the patenting of his spleen? And was his concept of self disturbed by the patenting of his body part? Moore's cancer-fighting cells are what make him unique. And in his mind, his patenting has been a violation of the self. He has been quoted as saying, "How does it feel to be patented? To learn all of a sudden, I was just a piece of material. . . . It's so beyond anything you can conceive of. There were so many issues involved. . . . There was a sense of betrayal. . . . I was told that, in a dinner conversation with a colleague, [my doctor] had said that, 'John Moore is my gold mine.' This was certainly true; he had discovered genetic gold in my cells."[87]

---

The pursuit of patentable genes has often been compared to the American gold rush.[88] Thousands upon thousands of scientists comb through the human genome in search of another genetic gold mine, just as gold diggers at one time rushed west and north to hack away at the land in hope of finding riches. Today's genetic prospectors search tentatively, almost blindly, through uncharted territory that may or may not be hiding a fortune. Everyone wants to be the first to find the treasure, because discovery, it seems, equates with ownership. Yet while the gold diggers could only claim ownership over what their hands could grasp, those mining for genetic gold claim property rights over the very process of discovery, as well as a monopoly on future unearthings of their source of biotech riches.

Genetic prospectors have set their sights on the developing countries. Thousands of miles from their laboratories, they are staking claim to the resources of indigenous people who have been taking advantage of nature's medicinal qualities for centuries. History, it seems, does not supersede a carefully drafted patent. The livelihoods of entire cultures are being exchanged for billions in revenue. For those Western corporations cashing their royalty cheques, it is a lucrative and industrious business plan. For others, it is biopiracy.

Chapter Nine

# The Gene Thieves

## Biopiracy in the Developing World

In 1991, a group of scientists announced that humanity was at a genetic crossroads. If the DNA of indigenous people was not preserved, it would be lost forever in the world's amalgamated gene pool. Our genetic roots would be lost, and our "pure" DNA blended beyond isolation. "The genetic diversity of people now living harbours the clues to the evolution of our species," these scientists claimed in a letter to *Genomics* magazine, "but the gate to preserve these clues is closing rapidly." Scientists interested in human genetics, they warned, must take advantage of this "vanishing opportunity to preserve a record of our genetic heritage."[1] The Human Genome Diversity Project was thus proposed. Approximately 700 samples would be drawn from "endangered species" including Australian aborigines, the Saami people of the Kalahari, Latin American Indians, and the Sami of northern Scandinavia. These indigenous groups, also referred to as "isolates of historic interest," would provide scientists the clues to our genetic origins as well as to the diseases that afflict us. White blood cells would be given new life as cell lines within a laboratory. This would keep DNA alive indefinitely, long past the lives of its indigenous owners.

Leading the HGDP was Luca Cavalli-Sforza, along with Mary-Claire King, Ken Kidd, and Charles Cantor, all of whom are renowned in the field of genetics. The estimated price tag of the HGDP was US$30 million over five years. The scientists were given seed money from the Human Genome Organization, the National Institutes of Health, the U.S. Department of Energy, and the National Science Foundation. The

HGDP was even more ambitious than the Human Genome Project (HGP). While the goal of the HGP was to sequence one set of genes to represent all of us, the HGDP focused on the genetic differences among diverse populations across the entire globe. Rather than analyzing one standard genome, the scientists would examine the diversity of humanity, because all of humanity must be studied in order for us to truly understand what makes us who we are. The goal was to create a DNA bank, or a so-called genetic museum, containing blood, saliva, and hair samples of indigenous peoples around the world. The HGP had shown that scientists had the resources for such a complex genetic study. But did people understand its importance? Would the scientists be able to get the project off the ground in time, before many of the proposed subjects became so integrated with the rest of the world that their genes were no longer useful? The HGDP scientists reasoned that "It would be tragically ironic if, during the same decade that biological tools for understanding our species were created, major opportunities for applying them were squandered."[2]

Many indigenous people, however, saw a different irony in the HGDP's proposed DNA bank. Rather than bridging the divide between developed and developing worlds, it would widen the disparity even further. The project was nothing more than a crucible for costly pharmaceuticals. Western scientists would study the genes of indigenous people in the developing world to offer new disease therapies that only the developed world could afford. The genetically "pure" indigenous populations would be examined and categorized in terms of their DNA. The world would learn how they were different from the rest of humanity. Maybe they harboured a predisposition to Huntington's disease, like the population of Venezuela's Lake Maracaibo. Or maybe they suffered from high blood pressure, like the Navajo. Indigenous people had already been categorized by their geography, culture, and language. Now they would be pigeonholed by their DNA. It was genetic profiling, critics proclaimed. It was just another form of discrimination. Yet this was the opposite of the intention of Cavalli-Sforza, who believed that genetics could help eradicate racism rather than exacerbate the problem. In the past he had worked with groups in Africa and had publicly debated against racism.

His intention was to look at the differences between individuals in order to uncover their overall similarities. According to the project's outline, illustrating the nature of these generally immaterial differences and overarching similarities "would help to combat the widespread popular fear and ignorance of human genetics and will make a significant contribution to the elimination of racism."[3]

In the past, individuals have been categorized by their phenotype (physical appearance), which includes hair and skin colour. But as Marc Feldman, one of the project's planners, points out, "[t]he closer in you go from what you see on the surface, the more unity there is." Another project member, Georgia Dunston, was quoted as saying, "After the Diversity Project, we won't have the luxury of drawing distinctions between one another based on skin pigmentation anymore."[4] We may look different on the outside, but on the inside we are more similar than we could have ever imagined.

Yet indigenous people around the world were not buying this promise. The HGDP was nicknamed the Vampire Project by activists, who denounced it as an experiment in genetic colonialism. Their blood would be siphoned for First World cures. They were being told that their DNA could help fight diseases that afflicted all of the world's peoples, that millions were to be spent studying populations that were dying of poverty and disease, but would those who donated their DNA be able to afford the therapies thereby facilitated? "Our lives and lands have been under continual assault for the past 500 years," states Debra Harry, a Northern Paiute activist, ". . . we've endured generations of outright war and genocidal policies. Now they've come to take our blood and tissues."[5] Activist groups began a very public protest against the HGDP, and became outraged when they learned of plans to commence the project as soon as possible, before populations assimilated or, in a few instances, died out. "The assumption that indigenous people will disappear and their cells will continue helping science for decades is very abhorrent to us," says Rodrigo Contreras of the World Council of Indigenous Peoples (WCIP).[6] In 1993, the Canada-based Rural Advancement Foundation International (RAFI), since renamed the Action Group on Erosion,

Technology, and Concentration (ETC Group), accused the HGDP of mining indigenous DNA for the benefit of the Western world.

The appropriation of biological resources without the consent of the people from whom those resources have been taken, or without an offer for them to share in the benefits or profits therefrom, is known as biopiracy. The ETC Group, which fights for the intellectual property rights of indigenous people, pointed to the possible biopiracy of DNA. Just as their traditional knowledge had been appropriated and profited from, so could their genes be. Their natural resources had been patented by pharmaceutical companies in the form of new medications. Would their very cells be harvested next? As Debra Harry states, "The assumptions posed by the HGD Project that the origins and/or migrations of indigenous populations can be 'discovered' and scientifically 'answered' is insulting to groups who already have strong cultural beliefs regarding their origins."[7] The HGDP was eventually disbanded, but "bioprospecting" for traditional knowledge and natural resources in the developing world, for the benefit of the developed world, continues to endanger indigenous life. This is far from a new phenomenon. Since the sixteenth century, scientists have ventured far from home in search of valuable natural resources that would bring them riches. The treasure hunt is for plants and animals that locals have identified as having medicinal qualities. Often with the help of locals, scientists identify valuable resources and learn how to release their healing powers. Sometimes these specimens are appropriated by foreign researchers who transport them back to the developed world.

Bioprospectors sift through the DNA and biochemicals of the developing world in pursuit of the next cash cow. A simple plant, if harvested and used correctly, can be worth millions in drug or pesticide sales. Take a mere soil sample from Norway's Hardangervidda National Park that was taken in 1969 by a researcher from Sandoz (now Novartis). Three years later, the immunosuppressant property of cyclosporine, an anti-rejection drug used for grafts and transplantations, was also taken from the soil. In 1983, Sandoz launched the drug commercially, and by 2000 it was the 33rd best seller across the globe, bringing in sales of US$1.2 billion. As researchers search more remote areas for yet more valuable extracts, they tread heavily on the toes of indigenous peoples. During the 1960s, bacteriologist and University of Wisconsin-Madison professor Thomas Brock found another precious resource, this one in Yellowstone National Park. Studying bacteria, Brock found *Thermus aquaticus,* which is named for its attraction to hot water. The bacteria's enzymes function at hot temperatures (unlike most enzymes from organisms), which is useful because chemical reactions happen at an accelerated rate in a hot environment. *T. aquaticus'* enzymes, it was discovered, could be used to convert massive amounts of corn into sugar for soft drinks. As well, the bacteria contains TAQ polymerase, which is used in chain reactions that replicate DNA unbelievably quickly. So huge amounts of a person's DNA can be replicated for genetic engineering or other biotech uses. It has been estimated that the enzyme brings in yearly sales of over US$200 million.[8]

Traditional knowledge has long been discovered among indigenous communities and its properties studied for potential commercial use. Take the neem plant, which originated in Southeast Asia and is now grown across the tropics for its medicinal and agricultural properties. The plant is used to treat flu, certain skin diseases, malaria, and even meningitis. It can also ward off insects and disease that destroy crops. Even though neem had been used by local indigenous communities for decades, a patent was granted in 1994 by the European Patent Office to the American company W.R. Grace and the U.S. Department of Agriculture for a "method for controlling fungi on plants by the aid of a hydrophobic extracted neem oil."[9] The following year, several international organizations, including some representing Indian farmers, opposed the patent. Neem did not exhibit novelty, they argued, because its fungicidal abilities had been used for centuries to protect crops in India. In 1999, the European Patent Office concluded that the patent was invalid because the use of neem was not a new invention. In 2000 the patent was revoked.

Shamans of Amazon tribes have long used the bark of *Banisteriopsis* to make the ceremonial drink *ayahuasca,* "vine of the soul." Ayahuasca is also relied on for its healing powers. An American citizen, Loren Miller, filed a patent application on the plant, saying he had discovered it growing in a domestic garden in the Amazon rain forest. Miller called it Da Vine, and claimed it was a new variety due to its flower colour. In 1986, Miller was granted exclusive rights to profit from the plant. In 1994, the Coordinating Body of Indigenous Organizations of the Amazon Basin (COICA), which represents over 400 indigenous populations, learned of the patent and asked for it to be re-examined. The Center for International Environmental Law, representing COICA, argued that Da Vine was neither "new" nor "distinct." In 1999, the patent claim was rejected. Yet in 2001, this decision was reversed and the patent was upheld.[10] For tribes of the Amazon, the ayahuasca plant is sacrosanct. "Some indigenous people say the patenting of this plant is the equivalent of somebody in their group patenting the Christian cross," explains David Rothschild of the Coalition for Amazonian Peoples and their Environment.[11] Just because some plants and animals have been unknown to the Western world does not mean that they were not familiar and useful in the developing world.

Although bioprospecting is often compared to prospecting for gold or oil, as author and activist Vandana Shiva points out, "this metaphor suggests that prior to prospecting, the resource lies buried, unknown, unused, and without value. Unlike gold or oil deposits, however, the uses and value of biodiversity are known by the communities from where the local knowledge is taken through bioprospecting contracts."[12] Generations of indigenous people have learned how to benefit from the plants and animals surrounding them. This is shared, traditional knowledge that is not owned or monopolized by one individual. And for indigenous people, this knowledge is a culturally distinct and sacred part of nature that exists to improve life itself, rather than the health and bank accounts of a select few.

---

Western scientists have now taken the concept of a patentable invention to another level, prospecting the genes of indigenous people and seeking exclusive rights to the "pure" cells within them, thanks to their geographical isolation. Scientists are fascinated by the medical possibilities, while shareholders are enamoured with the potential profits from genetics. In the early 1990s, some American scientists travelled to the rain forests of Panama, where over 120,000 Guaymi Indians live. These scientists collected blood, hair, and cell samples of the Guaymi, including specimens from one very interesting subject—a 26-year-old woman whose cells were thought to have unique antiviral qualities. She seemed to harbour a retrovirus typically associated with AIDS and leukemia, but did not show symptoms of either condition. A sample from her was used to create a cell line that contained her genetic code and would stay alive indefinitely in the laboratory, for future study.

In 1991, the U.S. Department of Commerce filed for a patent on the woman's cell line, without her knowledge or consent. According to Pat Mooney of the ETC Group, the Guaymi had no idea what their samples were to be used for, and did not understand the concept of a patent. Mooney stumbled across the patent application during an unrelated search of a patent database. He informed the World Council of Indigenous Peoples as well as the Guaymi General Congress. "To patent human material, to take human DNA and patent its products, that violates the integrity of life itself and our deepest sense of morality," said Isidro Acosta, president of the Guaymi General Congress, when he learned the news.[13]

While the patent application was withdrawn, the woman's cell line was not immediately returned, as the Guaymi people had requested. Samples continued to be sold to researchers for US$136. "Will profits be made from the genes of poor people whose physical survival is in question?" asked Mooney in the mid-1990s. "What benefits, if any, will accrue to the indigenous people from whom DNA samples will be taken? Couldn't the money being used to fund the project be better used to provide those same poor people with access to clean water, vaccinations and public health services?"[14]

In the developing world, biodiversity is the lifeblood of the people. The diversity of species and ecosystems is essential for medical, ceremonial, and agricultural purposes. It is sacred because it gives people their sense of identity within the world. The Philippines is renowned for its biodiversity, with over 40,000 species of wildlife and many varieties of forests, coral reefs, and ecosystems. According to the World Rainforest Movement, there are 14,500 plant types in the Philippines, with 5 percent of the world's known flora living here. Sadly, there are 192 wildlife species that are now endangered. And the destruction of the country's ecosystems is more severe than anywhere else in Southeast Asia, possibly anywhere else in the world.

The country's leaders are making insufficient efforts to correct the situation. The Philippine government has done little to deal with an ever-increasing population, nor has it emphasized sustainable development to safeguard its precious natural resources. Only 27 percent of the country's original mangrove forests are left standing, due to excessive logging and conversion of the forests into farmland. As a result, the organization Conservation International has declared the Philippines the world's conservation priority. In response, the Philippine government has implemented regulations on biodiversity prospecting, setting a global precedent. In 1997 the Philippines' Indigenous People's Rights Act was enacted to protect the culture and communities of indigenous populations. The Philippines is inhabited by approximately 4.5 million indigenous Filipinos from more than 70 "ethnolinguistic" groups, and the act is said to be the most powerful national law of its kind. In 2001 the Philippines' Wildlife Act became law. The purpose of these legislative initiatives is to protect the country's indigenous groups, along with its biological, environmental, and genetic resources.[15]

Nevertheless, bioprospectors do not always abide by rules and regulations protecting a country's biodiversity. The prospect of profit in the Western world continually threatens biodiversity in the developing world. "The ancestral tradition of sharing knowledge and freely exchanging seeds, plants and other resources—which has formed the very basis of diversity—may become a dangerous activity because once Indigenous Peoples share information or genetic resources with bioprospectors, it is possible they will lose control over those resources," states the ETC Group. "Given the majority of livelihoods in the South are dependent on biodiversity, losing control over their own genetic resources is one of the biggest threats to Indigenous Peoples and traditional communities."[16]

---

Biopiracy—bioprospecting for the sole purpose of making a profit, and a large profit, at that—has become a very big business. A pocketful of soil can result in a long-sought-after patent and a multi-million-dollar drug. Patents are filed, using indigenous medicinal knowledge of plants, animals, and biochemicals, and the local way of life is overturned. Once a patent is filed on, say, a type of rice, farmers who have been growing that particular rice for centuries no longer have the right to grow it. For example, basmati rice is traditionally grown in South Asia. Texas-based RiceTec was granted a patent on a type of basmati it had invented, using genetic material that was freely available. Indian farmers became outraged, accusing RiceTec of appropriating their traditional knowledge and the basmati name. Fortunately, the patent has been challenged and most of its claims revoked. If such a patent were to stand, farmers would have to pay for permission to use their own natural resources. According to the "Collective Statement of Indigenous Peoples on the Protection of Indigenous Knowledge," "Indigenous knowledge is the foundation of indigenous cultures. This knowledge permeates every aspect of our lives and is expressed in both tangible and intangible forms. Indigenous knowledge reflects the wisdom of our Ancestors, and we have a responsibility to protect and perpetuate this knowledge for the benefit of our future generations."[17]

While the term "biopiracy" was only recently coined, the concept is timeless. Over 3,500 years ago, Egyptians returned home from military expeditions with plants they had collected along the way. Charles Darwin, during his voyage on the *Beagle,* removed what he found fascinating from the Galapagos, bringing specimens home with him. England's Royal Botanical Gardens removed rubber trees from Brazil and gave them a new home in Southeast Asia. Even though it was illegal, the gardens appropriated cinchona seeds from Bolivia and planted them in India.

(Cinchona are roughly 40 species of flowering trees whose healing powers are used in herbal remedies for malaria, anemia, and indigestion.) More recently, the American scientific explorer of the Amazon Richard Schultes became acquainted with shamans who gave him permission to collect thousands of plant species with medicinal qualities that had not yet been recorded.[18]

Biopiracy has also been called biocolonialism. Westerners are invading the developing world, unearthing "inventions" to bring back to the developed world, announcing their personal discoveries of natural resources as if no one before them had comprehended their value. First it was theft of land and people; then came the appropriation of indigenous culture; now, traditional knowledge is being appropriated, patented, and profited from, with total disregard for the impact of commercialization on indigenous life, culturally or economically. According to an anthropologist, "less than 0.001 percent of the profits from drugs that originated from traditional medicines have ever gone to the indigenous peoples who led researchers to them."[19] Even though about 90 percent of worldwide biological resources originate in Asia and Africa, 97 percent of patents commercializing these resources are exercised by multinational corporations based in Western countries.[20] That is why Debra Harry is so critical of the developed world reaping benefits from the resources of developing countries. "The common thread," she says, is "that we're dealing with a white society that feels that anything that exists in indigenous territories is up for grabs."[21]

Yet bioprospecting, if done ethically, can be a valuable resource to both developed and developing nations alike. It can offer equal benefits to all participants. Biotechnology offers greater yields from agricultural crops, and treatments for the debilitating diseases of the developing world, while at the same time providing materials for Western industry. In the past, partnerships have been attempted between developed and developing countries. A novel benefit-sharing agreement between a country rich in biodiversity and a prosperous corporation involved pharmaceutical giant Merck & Co. and the Costa Rican National Biodiversity Institute (INBio). In 1989 the Costa Rican Ministry of Environment and Energy launched INBio with the goal of preserving the country's biodiversity, making a record of the valuable resources, and learning to use those resources productively as well as ethically. In 1991 INBio signed a historic deal with Merck. INBio agreed to provide the pharmaceutical company with chemical extracts from its plants, insects, and microorganisms for drug development. In return, Merck compensated INBio with US$1 million up front, as well as royalties on future drugs developed from the resources provided by INBio. The initial payment was used to set up INBio's own laboratory. Merck also provided equipment to run INBio's lab. There is a fee to use the lab's samples, which is used to pay for part of the operating costs. To help sustain Costa Rica's wealth of flora and fauna, 10 percent of the upfront payment and 50 percent of royalties earned by INBio will go to the government to fund conservation programs. To date there have been no royalty payments, as a drug has not yet been developed. Nevertheless, INBio continues to grow and prosper. It has 28 research labs and offices, and over 165 staff.[22] Other countries with similar diversity of plants and animals may also offer great financial opportunities. According to Vandana Shiva, the global market for medicinal plants identified by indigenous people is valued at US$43 billion. The seed industry alone is estimated to be worth US$7.5 billion.[23]

In the early 1990s, the Convention on Biological Diversity (CBD) was created to protect and preserve the world's biodiversity. The CBD asks individuals to "respect, preserve and maintain knowledge, innovations and practices of indigenous and local communities embodying traditional lifestyles relevant for the conservation and sustainable use of biodiversity."[24] The CBD maintains that nation-states have sovereign rights over their own biodiversity, and have the authority to control access to natural resources. This requires true informed consent and mutual terms of agreement between a country and visiting researchers. The CBD also acknowledges the market potential of biodiversity. There is potential development of technologies from biologically diverse countries that are "relevant to the conservation and sustainable use of biological diversity or make use of genetic resources and do not cause significant damage to the environment." If resources are used in a sustainable manner, they can yield productive crops and future products while conserving the environment.

The CBD was also created to ensure the "fair and equitable" sharing of benefits from genetic resources. Benefits can be monetary or non-monetary, such as free distribution of future pharmaceuticals, training costs, or a budget to launch or improve a laboratory within the host country. They can also vary in time frame: short-term benefits like cash are derived at the time access is granted; medium-term benefits such as funding for research or contributions to a local economy are awarded between initial access and the end result of commercialization of resources; and long-term benefits such as joint ownership of patents or a share in royalties are enjoyed once a product has been developed successfully.[25] If resources are shared equally between bioprospectors and original "owners," bioprospecting can lead to a profitable and sustainable industry for the future. Access to indigenous land, plants, and species can be exchanged for present and prospective benefits, helping to promote economies and biodiversity in the developing world.

---

The CBD states that access to resources should be in harmony with the "adequate and effective protection of intellectual property rights," and that conditions of access should be mutually agreed upon by all parties.[26] But intellectual property rights (IPR) governing the activity of commercial research in developing countries do not always make benefit-sharing a priority. IPRs are private rights that reward their owner with a monopoly on a product or process for a set period of time and the power to ban the competition from using the product or process. An IPR holder may decide to grant others use of the protected product, but this is a matter of choice. The IPR holder may decide to exclude some parties while allowing others to use the product, and may charge a fee for this privilege. The priority of IPRs, then, is commercialization, and "IPRs will affect who shares in the benefits arising from genetic resources, and the type of technology developed from genetic resources, with implications for the conservation and use of biological diversity."[27]

The British government decided to ascertain the impact that IPRs have on developing countries. It set up the Commission on Intellectual Property Rights (CIPR), which concluded that the patent system of developed countries can work in certain cultural contexts, while suppressing innovation and competition in others. The best possibility that can be hoped for, stated a 2002 CIPR report, is that Western patents encourage scientific advances some of the time.[28] Unfortunately, it is not the needs of the developing world that determine the nature of the patent system. Rather, this system is shaped by large multinational corporations that create jobs and fuel economies. If the research and development of big pharmaceutical and biotech companies are not protected under patent law, it is unlikely that such companies will invest in new medicines or technologies that advance the quality of life in both the developed and developing worlds. As noted earlier, patents help companies cover the long drug development process by letting them use their monopoly to maximize profits and keep the competition at bay.

In 1994 the World Trade Organization (WTO), an international association of 142 countries that examines trade among nations, drafted "Trade-Related Aspects of Intellectual Property Rights" (TRIPS) to determine harmonized standards for the protection and implementation of intellectual property rights of patents, copyright, trade secrets, and trademarks. These standards are the minimum that must be enforced by member countries, rather than the maximum allowed; members can choose to implement broader protection. The goal of TRIPS was to "reduce distortions and impediments to international trade" while encouraging technological advances in ways "conducive to social and economic welfare, and to a balance of rights and obligations."[29] Developed member countries had to adopt TRIPS regulations by 1996, developing countries were given until 2000, and the least-developed countries have until 2006. Products in development before January 1, 1995, are exempt from the regulations.

While TRIPS was intended to provide incentives to create new technology and products to improve life, many observers fear that it will have the opposite effect. Rather than minimizing inequalities between the developed and developing world, international trade could in fact widen disparities among peoples. Even though advances in medicine have resulted in treatments for most of the major infectious diseases, including HIV/AIDS, malaria, and tuberculosis, the benefits from such scientific innovation are not universally distributed. The World Health Organization (WHO) estimates that one-third of the world's people lack access to essential medications, with over 50 percent of people in Africa's and Asia's poorest areas lacking the most basic, essential drugs. Essential medications, by WHO definition, are "those which satisfy the health care needs of the majority of the population, should be affordable, and represent the best balance of quality, safety, efficacy and cost for a given health setting."[30]

In developing countries, between 25 percent and 65 percent of the health budget is spent on pharmaceuticals. Yet this still isn't enough to provide medication for everyone, and many cannot afford to pay for their own treatment. According to the WHO and WTO, access to such drugs requires sustainable financing, a reliable system of supply, and rational selection and use of medications, as well as affordable pricing.[31] Under TRIPS, patent protection on a development process as well as on a product itself was extended to 20 years. Before TRIPS, patents were issued on processes only, allowing countries to use different methods to create cheaper generic versions of patented medications. Now, such production of generics is banned until the patent protection expires. When TRIPS was first enacted, approximately 50 member countries did not give patent protection for pharmaceuticals. Soon they will all have to abide by patent regulations, further decreasing the Third World's already inadequate access to drugs.

A number of exceptions to the patent rules have been written into TRIPS in order to aid the poor and the sick in developing countries. Article 8.1 of TRIPS allows for countries in crisis to "adopt measures necessary to protect public health and nutrition, and to promote the public interest in sectors of vital importance to their socio-economic and technological development."[32] For countries with desperately ill people who cannot afford patented medication, TRIPS allows companies to offer "differential pricing," also referred to as "tiered" or "equity" pricing. This is a controlled pricing system for critical drugs, whereby the price of a pharmaceutical remains relatively high in developed countries, so as not to diminish the company's R&D budget, but is lower in developing countries. However, voluntary price cuts implemented by many drug firms to reduce the cost of treatment for HIV/AIDS by 90 percent did not eliminate the problem of access. Many of the world's poorest countries are still unable to afford this medication. Some drug companies give donations to countries in need; while this alleviates the immediate crisis, it does not help the countries prepare for the future, once the money has been spent.

The TRIPS exception of "parallel importing" occurs when drugs are bought through a third party in another country rather than directly from the manufacturer, to avoid the price the drug company has set for the purchasing country. Britain often uses parallel importing so that its citizens get the best possible prices. In the United Kingdom, for example, Retrovir, an anti-HIV treatment, is priced at £125. Yet people can buy the same drug imported via other countries in Europe for £54.[33] The variation in pricing can be attributed to differences in local incomes, and competition among pharmaceutical producers. According to a 1998 study by the Consumer Project on Technology, prices for the antibiotic Amoxil vary greatly. It was selling at US$8 in Pakistan, US$14 in Canada, US$40 in Indonesia, and US$60 in Germany.[34]

Under TRIPS, "compulsory licensing" permits countries to manufacture and use generic versions of drugs without the permission of the patent owner under certain circumstances. A national judicial body can grant such a licence if a developing country can show that a new medication, such as an AIDS vaccine, could relieve a public health crisis, or if it discovers anti-competitive practices on the part of a drug manufacturer. Compensation is granted to the patent holder under such rules, as determined by the law of the country that grants the licence. The difficulty with compulsory licensing is that many developing countries do not possess the infrastructure to manufacture drugs themselves, and most of their populations cannot afford to buy even cheap generics. And although

TRIPS makes compulsory licensing legal, it does not necessarily follow that big pharmaceutical companies will agree.

Big pharma has formulated its own alternative set of regulations, called "TRIPS-PLUS." As author Jeffrey Robinson explains, pharmaceutical companies "warn governments and health ministries that it is illegal for generic manufacturers to develop drugs in a country when a valid patent is still in force (which is not true everywhere); that the practice of compulsory licensing, which allows poor countries to produce essential drugs in a national emergency, is flat-out illegal (untrue); and that no national health system is permitted to use only the generic product when the brand-name drug is available on the market (again untrue)."[35] Robinson cites the example of Thailand, which at one time had only one medication for cryptococcal meningitis, a fatal disease that often strikes AIDS patients. Pfizer offered to supply the drug Triflucan, but at a price the average Thai patient could not even attempt to afford; for one month, the cost was more than $400. Then a pair of companies in Thailand decided to develop a generic version of the drug and sell it for less than a third as much. Many patients could still not afford the treatment, but others could. However, Pfizer was losing money, and the U.S. government intervened. If the generic manufacturers didn't halt their business, writes Robinson, Washington would tax Thai exports to the U.S. "[Thailand] had a statute in place that gave them the authority to [manufacture generic drugs], and it was consistent with international law," says James Love of Washington's Consumer Project on Technology. "But the U.S. government threatened trade sanctions, and used a carrot and stick approach to persuade the Thai government not to do something which would have been legal under international law."[36] TRIPS-PLUS is so named, says Robinson, because it is what big pharma would like to add to TRIPS. But it is nothing more than a wish list, and is not part of any patent law.

Thailand is not the only country that has used compulsory licensing to challenge the U.S. drug industry. In developed countries, the use of antiretroviral (ARV) therapies has transformed AIDS from a fatal to a chronic illness for many patients. Yet ARVs are extremely expensive, costing between US$10,000 and US$17,000 annually. In developing countries, where such treatment is far beyond almost everyone's means, AIDS remains a terminal disease, with the average life expectancy between seven and nine years once HIV is detected.[37] South Africa is home to four million AIDS sufferers.[38] As a South African judge infected with HIV explained in 2001, "International agencies, national governments, and especially those who have the primary power to remedy the inequity—the international drug companies—have failed us in the quest for accessible treatment."[39]

In 1997 Nelson Mandela's government passed the Medicines and Related Substance Control Amendment Act, aiming to make drugs affordable to the South African people. The legislation encompassed three key initiatives: the substitution of generics for expensive brand-name medication by pharmacists; parallel importing of patented medications from countries where the company sold its drugs at a reduced cost; and the creation of a pricing committee responsible for ensuring a fair pricing mechanism. According to the "National Drug Policy for South Africa," the act was created "to ensure an adequate and reliable supply of safe, cost-effective drugs of acceptable quality to all citizens of South Africa and rational use of drugs by prescribers, dispensers and consumers."[40] This did not go over well with the U.S. government.

In 1997, the Clinton administration informed the South African government that if it did not rethink its policy, the U.S. would impose trade sanctions. According to the U.S. Trade Act of 1974[41] and its 1988 updates, the United States can enforce economic sanctions against countries that do not abide by proper trade regulations. If a nation abides by TRIPS but, according to the U.S., is not offering "fair and adequate" patent protection, sanctions may be enforced.[42] In 1998, South Africa was put on a U.S. trade watch list. That year, the country was deprived of Generalized System of Preferences (GSP) treatment, which allows products to enter the U.S. duty-free.

Thirty-nine pharmaceutical companies sued the South African government, alleging that the Medicines and Related Substance Control Amendment Act contravened the country's constitution. South African representatives explained that the legislation did not include compulsory licensing, and did not grant powers to overrule the interests of patent holders, and so did not conflict with the constitution. The multinationals stated that the act also went against the TRIPS agreement, while the South African government insisted that it was consistent with TRIPS regulations enabling WTO members to authorize parallel importing. "There was a feeling that if a country deliberately went against TRIPS, there would be a castle-of-cards effect," said Jean-Pierre Garnier, chief executive officer of pharmaceutical giant GlaxoSmithKline. "Without patents, the industry ceases to exist."[43] However, a case was never formally brought to the WTO accusing South Africa of violating TRIPS.[44]

In 1999, Charlene Barshefsky, U.S. trade representative under the Clinton administration, called for a review of South Africa based on its advocacy role in the World Health Assembly. In January 1999, at the 52nd World Health Assembly, a resolution had made health a part of trade negotiations. The U.S. backed down from its harsh stance on South Africa, removing it from its trade watch list in 2000. Thousands around the world protested the big pharma case regarding the South African constitution, saying the country was "suffering an apartheid of drugs" from stringent patent laws. "Here in the wealthy West, we have antiretroviral drug cocktails which prolong life, improve the quality of life, and serve, as it were, to save life," Stephen Lewis, the Secretary-General's special envoy for HIV/AIDS in Africa, was quoted as saying. "We have the drugs. We use them. In the developing world, where 95 percent of the new infections occur, virtually everyone HIV-positive is doomed to a gruesome and painful death. The numbers of people who can afford the drug cocktails are so infinitesimal as to be invisible."[45]

In 2001, the pharmaceutical companies dropped the case in the High Court of Pretoria, agreeing to pay South Africa's legal fees. During the proceedings, the South African government had chosen to hold off on implementing the Medicines and Related Substance Control Amendment Act for fear of trade retaliation by the U.S. government. According to the Canadian HIV/AIDS Legal Network, during the three years that the act was suspended, over 400,000 South Africans died of AIDS.[46]

The reality is that developing countries have never been a priority for pharmaceutical companies. Producing drugs for the world's poorest nations is simply not cost-effective. Take Africa as an example. While the global drug market is worth over US$400 billion a year, Africa accounts for a mere 1.3 percent of this market.[47] The disparity of health spending between the developed and developing worlds was coined the "10/90 gap" by the Global Forum for Health Research in 2000; of the US$50 billion to US$60 billion spent worldwide on health research by private and public institutions each year, only 10 percent is allocated to the health problems of 90 percent of the world's people.[48] There is just not enough money in this 90 percent.

In order for drug companies to survive, they must produce enough "blockbuster" drugs to reap a profit, which then funds the development of other products still in the pipeline. A typical blockbuster drug exceeds $1 billion in revenue annually for its company. The world's so-called neglected diseases are not usually targets for blockbusters, because most people who suffer from them do not have the money to buy the drugs anyway. In many poor countries whose economies are agriculture-based, US$22 per capita is spent annually on health. This drastically low figure includes money for drugs and hospitals.[49] Blockbuster treatments are for typically Western conditions such as heart disease, cancer, and diabetes.

According to the WHO, pneumonia, diarrhea, tuberculosis, and malaria account for more than 20 percent of the world's burden of disease, and receive less than 1 percent of all public and private money allocated for health research. In 1998, only US$300 million was allotted to vaccines for HIV/AIDS, and US$100 million to research on malaria. And of the 1,233 medications that hit the market between 1975 and 1999, only 13 specifically targeted tropical diseases. Of these, six were developed by the WHO, United Nations Development Programme (UNDP), and the UNDP/World Bank WHO–supported Special Programme for Research and Training in Tropical Diseases (TDR).[50] It is easy to understand why life expectancy in developed countries ranges from 75 to 80 years of age, while in the world's least developed countries it ranges from 40 to 50 years.[51]

While the extraordinary potential of medicine based on genomics promises to improve life in the developed world, there is a danger that complex, costly technologies may worsen the already ailing health of developing nations. Regenerative medicine may widen the gap between the West and the Third World underclass, who are unable to benefit from stem cell research, gene therapy, or the mapping of the genome. Medical advancements are costly, not only to develop but to use. Fortunately, much can be done to combat this health divide. Developed countries can form strong partnerships with their developing neighbours, offering fair benefit-sharing for the use of indigenous resources. This should not come only as immediate cash relief; to lengthen their population's life expectancy in the future, developing countries must learn how to be self-sufficient, implementing genetic programs for their own people.

With help from more developed nations, Third World doctors and researchers can bootstrap themselves into the age of genomics. Given time and support, they too can use the techniques and equipment of DNA-based diagnostic testing to scan patients for genetic predispositions, as well as infectious diseases; they too can customize therapies according to the person's individual drug resistance. Already, some striking progress has been made.

San Francisco's Sustainable Science Institute has created north/south partnerships to train people in developing countries to use DNA diagnostic equipment. The goal is to create a cost-effective method of disease detection in the developing world. In Nicaragua, DNA diagnosis is being used for rapid identification of dengue hemorrhagic fever, leptospirosis, and leishmaniasis. Tanzania has developed a program with the Swiss Tropical Institute to identify drug resistance in patients with malaria, and to genotype malarial parasites. Using genetics, Africa is identifying resistance to certain HIV/AIDS drugs.[52] Genetics partnerships between the developing and developed worlds will succeed if they are based on mutual respect and equality. Every year, scientists and physicians learn more about the body's regenerative capabilities, using what is within each of us—our stem cells, genes, and proteins—to ward off the diseases that threaten our lives. The cells within us truly are golden; we are our most precious resource. The restorative powers of our cells continue to be discovered, improving and extending life.

The future holds unbelievable promise for regenerative medicine. In 2004, University of Toronto microbiologist Derek van der Kooy isolated stem cells in the human eye. This is a tremendous discovery, considering that the eye does not regenerate. After retinal stem cells produce the other types of cells in the eye while it is forming in the womb, they become dormant. In June 2005, van der Kooy said he is hopeful that clinical trials using stem cells to restore vision will begin in five years. One retinal stem cell could create 10,000 new cells in seven days. The goal is to remove retinal stem cells in blind patients and grow the types of required eye cells in the lab to repair blindness.[53]

In June 2005, Dr. Andras Nagy and a team of researchers at Mount Sinai Hospital in Toronto created Canada's first pair of human embryonic stem cell lines. Many of the stem cell lines in existence are outdated, or come with restrictions as to how they can be used. Canada is one of a few countries in the world to have its own stem cell lines to advance regenerative research.[54]

In March 2005, American researchers reported that stem cells found in hair follicles have the ability to develop into nerve cells, possibly offering a new source of cells to treat Parkinson's disease or spinal cord injuries. Gene therapy is now being hailed as a potential future treatment for Alzheimer's disease. In one study that concluded in the spring of 2005, skin cells were taken from eight patients and genetically modified to secrete a protein that is naturally found in healthy brains, which is known as nerve growth factor (NGF). A team of surgeons from the University of California injected the NGF-producing cells into patients' brain tissue damaged by Alzheimer's. For two years, the patients were observed. The resulting memory tests revealed that cognitive decline had slowed by as much as 50 percent, and scans demonstrated that the patients' brains had become more active than they had been before the gene therapy. The physicians involved in this novel treatment stress that it isn't a cure and much work remains to be done. But the initial results are inspiring for the 4.5 million Americans and

420,000 Canadians over age 65 who have Alzheimer's and related dementias.

---

Regardless of geography, race, or class, it seems that we are all bound by our genes. We may remain strangers to those living hundreds of miles from the place we call home, but our genetic blueprint is our common denominator. Regenerative medicine will use what is within all of us to cure the diseases that threaten our very lives, today or tomorrow. We house our own regenerative tool kits to bring us back to health. Our genome offers us road maps to our diseases. Our faulty genes can be reprogrammed to correctly produce the proteins that do everything from digest our food to combat sickness. Our stem cells can mend broken hearts, empower weakened muscles, and bring delusional minds back to reality. Regenerative medicine holds the power and promise of unlocking what is within all of us—life.

# Notes

## Chapter One: Taxicabs and Roadblocks

1 Description of surgery and quotations from phone interview with Nathan Klein.
2 Description of Dr. Michael Kaplitt's childhood and quotations from phone interview with Dr. Martin Kaplitt.
3 Quotations and explanations of the surgical procedure from phone interviews with Dr. Michael Kaplitt.
4 Sheryl Gay Stolberg, "The Biotech Death of Jesse Gelsinger," *The New York Times Magazine* (November 28, 1999): 137.
5 Information on viruses from phone interviews with Dr. Michael Kaplitt.
6 Parkinson's Society of Canada statistics for 2003.
7 For an excellent description of Parkinson's disease and its treatments see Dr. Michael Kaplitt's website at www.beatparkinsons.com.
8 Description of how a virus is prepared for surgery as well as the surgery itself from phone interviews with Dr. Michael Kaplitt.
9 Phone interview with Dr. Martin Kaplitt.
10 Barbara Sibbald, "Death But One Unintended Consequence of Gene-Therapy Trial," *Canadian Medical Association Journal* 164:11 (May 29, 2001): 1612.
11 Description of car accident and quotation from Gelsinger from "Jesse's Intent," a speech written by Paul Gelsinger.
12 Stolberg, "The Biotech Death of Jesse Gelsinger": 138.
13 Ibid: 139.
14 Ibid.
15 Deborah Nelson and Rick Weiss, "Penn Researchers Sued in Gene Therapy Death: Teen's Parents Also Name Ethicist As Defendant," *Washington Post* (September 19, 2000): A3.
16 Phone interview with Paul Gelsinger.
17 Quotations and description of gene therapy procedure from "Jesse's Intent."
18 Stolberg, "The Biotech Death of Jesse Gelsinger": 149.
19 Phone interview with Paul Gelsinger.

20 Information regarding meeting between Dr. Robert Erickson, Dr. Steven Raper, and Paul Gelsinger from phone interview with Paul Gelsinger. For Dr. Robert Erickson's apprehensions see also Stolberg, "The Biotech Death of Jesse Gelsinger": 139.
21 Eliot Marshall, "Clinical Trials: Gene Therapy Death Prompts Review of Adenovirus Vector," *Science* 286:5448 (December 17, 1999): 2244.
22 Rick Weiss and Deborah Nelson, "Methods Faulted in Fatal Gene Therapy," *Washington Post* (December 8, 1999): A1.
23 Ibid.
24 Stolberg, "The Biotech Death of Jesse Gelsinger": 150.
25 Quotation and information about the Gelsinger trial from phone interview with Alan Milstein.
26 International Conference on Harmonisation of Technical Requirements for Registration of Pharmaceuticals for Human Use, "ICH Harmonised Tripartite Guideline: Guideline for Good Clinical Practice" (May 1, 1996) 9. Online: ICH Homepage www.ich.org.
27 University of Vermont, The Community Genetics and Ethics Project, "Ask CGEP." Online: University of Vermont Homepage www.uvm.edu/~cgep/Education/Expert.html.
28 "Bioethics: Gene Therapy Business: the Tragic Case of Jesse Gelsinger," *News Weekly* (August 12, 2000). Online: Newsweekly Home Page www.newsweekly.com.au/articles/2000aug12_bio.html.
29 International Conference on Harmonisation of Technical Requirements for Registration of Pharmaceuticals for Human Use, "ICH Harmonised Tripartite Guideline": 15.
30 Tom Hollon, "Researchers and Regulators Reflect on First Gene Therapy Death," *Nature Medicine* 6:1 (January 2000): 6.
31 Jennifer Washburn, "Informed Consent: Alan Milstein says he wants to rescue us from unscrupulous doctors, undisclosed risks and greedy institutions. But is he a shining knight, or an enemy of medical progress?" *Washington Post* (December 30, 2001): W16.
32 From telephone interview with Dr. Anthony Ridgway.
33 Department of Health and Human Services, National Institutes of Health, Recombinant DNA Advisory Committee, Minutes of Meeting, June 8–9, 1995. Online: NIH Homepage www4.od.nih.gov/oba/rac/minutes/6-8-9-95.htm.
34 Association of University Technology Managers, "AUTM Licensing Survey: FY 2002. A Survey Summary of Technology Licensing and Related Performance for U.S. and Canadian Academic and Nonprofit Institutions, and Patent Management and Investment Firms": 1–2. Online: AUTM Homepage www.autm.net.
35 U.S.C. 200–212.
36 Nancy Forbes, "Managing Conflicts of Interest," *The Industrial Physicist* (August/September 2001): 22.
37 Business-Higher Education Forum, "Working Together, Creating Knowledge:

The University-Industry Research Collaboration Initiative" (2001): 19–22. Online: BHEF Homepage www.bhef.com.

38 Business–Higher Education Forum, "Working Together": 19.

39 Tadahiro Ohkoshi, *Asia Pacific Perspectives: Japan+* (September 2003). Online: www.jijigaho.orljp/app/0309/eng/sp_wark02.html.

40 World Medical Association, Declaration of Helsinki: "Ethical Principles for Medical Research Involving Human Subjects" (June 1964) Part B, Section 22. Online: WMA Homepage www.wma.net/e/policy/b3.htm.

41 Business-Higher Education Forum, "Working Together": 38.

42 American Association of University Professors, "Statement on Corporate Funding of Academic Research." Online: AAUP Homepage www.aaup.org/statements/Redbook/repcorf.htm.

43 Medical Research Council of Canada, Natural Sciences and Engineering Research Council of Canada, Social Sciences and Humanities Research Council of Canada, "Tri-Council Policy Statement: Ethical Conduct for Research Involving Humans" (1998, with 2000 and 2002 updates): 4.1. Online: Interagency Advisory Panel On Research Ethics Homepage www.pre.ethics.gc.ca.

44 W. French Anderson, "Genetics and Human Malleability," *The Hastings Center Report* 20:1 (January/February 1990): 23.

45 Descriptions of Dr. W. French Anderson and quotations, see Leon Jaroff, "Battler for Gene Therapy," *Time* 143:3 (January 17, 1994): 56.

46 Descriptions of Dr. W. French Anderson and quotations, see Jaroff, "Battler for Gene Therapy": 56–57.

47 W. French Anderson, "The Best of Times, the Worst of Times," *Science* 288:5466 (April 28, 2000): 629.

48 Zina Moukheiber, "Science for Sale: Corporate Funding for Academic Labs," *Forbes* (May 17, 1999). Online: Forbes Homepage forbes.com.

49 Moukheiber, "Science for Sale: Corporate Funding for Academic Labs."

50 University of Vermont, The Community Genetics and Ethics Project.

51 Washburn, "Informed Consent: W16.

52 Deborah Nelson and Rick Weiss, "Penn Settles Gene Therapy Suit," *Washington Post* (November 4, 2000): A4–A5.

53 Deborah Nelson and Rick Weiss, "Gene Researchers Admit Mistakes, Deny Liability," *Washington Post* (February 15, 2000): A3.

54 FDA warning letter dated February 8, 2002, addressed to Dr. James M. Wilson, from Dennis Baker, Associate Commissioner for Regulatory Affairs. Online: FDA Homepage www.fda.gov/foi/nooh/Wilson.htm.

55 Harvey Black, "Wilson Leaves UPenn's Gene Therapy Institute: Director oversaw experiment that led to death of 18-year-old patient in 1999," *The Scientist* (April 23, 2002). Online: www.the-scientist.com /2002/4.

56 FDA and NIH press offices, "New Initiatives to Protect Participants in Gene Therapy Trials," press release (March 7, 2000). Online: NIH Homepage www.grants.nih.gov/grants/policy/gene_therapy_20000307.htm.

## Notes

### Chapter Two: The Gene Hunters

1 For descriptions of Leonore Wexler see Alice Wexler, *Mapping Fate: A Memoir of Family, Risk, and Genetic Research* (New York: University of California Press, 1996): 77–79.

2 Wexler cited in Jeff Lyon and Peter Gorner, *Altered Fates: Gene Therapy and the Retooling of Human Life* (New York: W. W. Norton & Co., 1995): 415.

3 Robert F. Mueller and Ian D. Young, *Emery's Elements of Medical Genetics*, 11th ed. (London: Churchill Livingstone, 2001): 3–4.

4 Ibid: 5

5 James D. Watson, *DNA: The Secret of Life* (New York: Alfred A Knopf, 2003): xi–xiv, 51–55.

6 See Eric S. Grace, *Biotechnology Unzipped: Promises and Realities* (Toronto: Trifolium Books Inc., 1997): 21–29.

7 Industry Canada. "Overview—What is Gene Therapy?" Online: Industry Canada Homepage www.strategis.gc.ca.

8 Descriptions of single-gene and complex-gene diseases from interview with Dr. Roderick McInnes.

9 Matt Ridley, *Genome: The Autobiography of a Species in 23 Chapters* (New York: HarperCollins, 1999): 55.

10 Ibid.: 56.

11 For descriptions of RFLP mapping see Lyon and Gorner, *Altered Fates*: 416–417.

12 Nancy S. Wexler, "Clairvoyance and Caution: Repercussions from the Human Genome Project," in *Code of Codes: Scientific and Social Issues in the Human Genome Project*, ed. Daniel J. Kevles and Leroy Hood (Cambridge: Harvard University Press, 1992): 211–43.

13 Ibid.: 216.

14 Christina Hoag, "A Tale of Pain and Hope on Lake Maracaibo," *BusinessWeek* (June 5, 2000): 20E10.

15 Lyon and Gorner, *Altered Fates*: 417–18.

16 Wexler cited in Lyon and Gorner, *Altered Fates*: 418.

17 Wexler, "Clairvoyance and Caution": 219.

18 Ibid.

19 Gusella cited in Lyon and Gorner, *Altered Fates*: 410.

20 Wexler, "Clairvoyance and Caution": 211–43.

21 Andrew Revkin, "Hunting Down Huntington's: From the Shores of Maracaibo to the Halls of Washington, Nancy Wexler Has Spent 25 Years Stalking Her Mother's Killer," *Discover* 14:12 (December 1993): 108.

22 Ibid.

23 For descriptions of finding the gene and testing the repeat see Lyon and Gorner, *Altered Fates*: 423–27.

24 Wexler cited in Andrew Revkin, "Hunting Down Huntington's": 98.

25 Nancy S. Wexler, "Will the Circle Be Unbroken? Sterilizing the Genetically

Impaired," in *Genetics and the Law II*, ed. Aubrey Milunsky and George J. Annas (New York: Plenum Press, 1980): 228.

26 Ibid.: 229.

27 Peter Crittendon, "Ending a Historical Taboo: Restoring the Respectability of Eugenics," *American Renaissance* 8:2 (February 1997). Online: American Renaissance Homepage www.amren.com/972issue/972issue.html.

28 Edwin Black, *War against the Weak: Eugenics and America's Campaign to Create a Master Race* (New York: Four Walls Eight Windows, 2003): 67–68.

29 Karl Pearson cited in Edwin Black, *War against the Weak*: 27.

30 Wexler, "Will the Circle Be Unbroken?: 231-32.

31 Ibid.: 235–236.

32 Mary Nemeth, "Nobody Has the Right to Play God," *Maclean's* 108:26 (June 26, 1995): 17.

33 "Professor Says Canada Sliding into Eugenics," *The Globe and Mail* (March 14, 2004). Online: Globe and Mail Homepage www.theglobeandmail.com.

34 Wexler, "Will the Circle Be Unbroken?: 236.

35 Black, *War against the Weak*: 317.

36 Ibid.: 318.

37 Daniel J. Kevles, "Out of Eugenics: The Historical Politics of the Human Genome," in *Code of Codes*: 3–36.

38 Black, *War against the Weak*: 422–25.

39 Christine Mlot, "Panel Backs Widening Net of Genetic Test," *Science News* 151:17 (April 26, 1997): 253.

40 David Concar, "Test Blunders Risk Needless Abortions: Two Years after the U.S. Began Nationwide Screening for Cystic Fibrosis, Confusion over the Test Results May Mean That Healthy Fetuses Are Being Terminated," *New Scientist* 178:2393 (May 3, 2003): 4.

41 Ontario Provincial Advisory Committee on New Predictive Genetic Technologies, *Genetic Services in Ontario: Mapping the Future* (Nov. 30, 2001): 9. Online: Ontario Ministry of Health and Long-Term Care Homepage www.health.gov.on.ca.

42 Ibid.: 10

43 Ibid.: 8; see also Peter Harper, *Practical Genetic Counselling* (Oxford: Butterworth and Heineman, 1998): 11.

44 Abby Lippman, "Prenatal Genetic Testing and Screening: Constructing Needs and Reinforcing Inequities," *American Journal of Law & Medicine*. 17:1, 2 (1991): 27.

45 For descriptions of different genetic testing see National Human Genome Research Institute, "What Is Genetic Testing?" and "Reasons for Genetic Testing." Online: National Human Genome Research Institute Homepage www.genome.gov.

46 Organization for Economic Co-operation and Development (OECD), *Genetic*

*Testing: Policy Issues for the New Millennium* (2000): 9 and 39. Online: OECD Homepage www.oecd.org.

47 Lippman, "Prenatal Genetic Testing and Screening": 19

48 Rosemary Quigley, "Skirmishes: A tour of duty on the genetic battlefields of cystic fibrosis," *Boston Review* (December 1998/January/1999). Online: www.bostonreview.net/BR23.6/quigley.html.

49 Canadian Press, "Professor Says Canada Sliding into Eugenics" (March 14, 2004).

50 C. Cameron and R. Williamson, "Is There an Ethical Difference between Preimplantation Genetic Diagnosis and Abortion?", *Journal of Medical Ethics* 29:2 (April 2003): 91.

51 Ibid.

52 Margaret A. Somerville cited in Marina Jimenez, "'Wrongful Life' Ruling Outrages Ethicists: Court Decision in France: Down Syndrome Children Can Sue MDs for Not Aborting," *National Post* (December 15, 2001): A1.

53 Dr. Roger Bessis cited in Jimenez, "'Wrongful Life' Ruling Outrages Ethicists": A1.

54 Patrik S. Florencio and Mark K. Searl, "Genetic Interventions in Minors and Designer Babies," *Health Law in Canada* 21:2 (November 2000): 51–52.

55 Ibid.: 52.

56 World Health Organization (WHO) Advisory Committee on Health Research, *Genomics and World Health* (2002): 161. Online: WHO Homepage www.who.int/genomics.

57 John Robertson, "Sex Selection: Final Word from the ASRM Ethics Committee on the Use of PGD," *The Hastings Center Report* 32:2 (March–April 2002): 6.

58 *Winnipeg Child and Family Services (Northwest Area) v D.F.G.*, [1997] 3 SCR. 925 at 939.

59 Florencio and Searl, "Genetic Interventions in Minors and Designer Babies": 53.

60 For statistics and descriptions of Rabbi Ekstein's program see Steve Jones, *In the Blood: God, Genes and Destiny* (New York: HarperCollins, 1996): 75–77.

61 OECD, *Genetic Testing*: 9, 33.

62 Nuffield Council on Bioethics, *Genetic Screening: Ethical Issues* (December 1993): 55. Online: Nuffield Council on Bioethics Homepage www.nuffieldbioethics.org.

63 Patrik S. Florencio and Erik D. Ramanathan, "Secret Code: The Need for Enhanced Privacy Protections in the United States and Canada to Prevent Employment Discrimination Based on Genetic and Health Information," *Osgoode Hall Law Journal* 39:1 (2001): 86.

64 Ruth Hubbard and Eliiah Wald, *Exploding the Gene Myth: How Genetic Information Is Produced and Manipulated by Scientists, Physicians, Employers, Insurance Companies, Educators, and Law Enforcers* (Boston: Beacon Press, 1999): 131–133.

65 Florencio and Ramanathan, "Secret Code": 81.

66 Ibid.: 82.

67 For descriptions of the three tenets of privacy see Florencio and Ramanathan, "Secret Code": 88–91.

68 Ibid.: 101.

69 Ibid.: 103. See *R. v Plant,* [1993] 3 SCR 281.

70 S.C. 2000, c. 5.

71 Florencio and Ramanathan, "Secret Code": 100–107.

72 Henry Greely, "Health Insurance: Employment Discrimination and the Genetics Revolution," in *Code of Codes*: 266.

73 Ibid.

74 Pub. L. 104–191, 110 Stat. 1936 (1996) (short title, see 42 U.S.C. 201 note).

75 OECD, *Genetic Testing*: 51-53.

76 42 U.S.C. 12101.

77 Anita Silvers and Michael Ashley Stein, "Human Rights and Genetic Discrimination: Protecting Genomics' Promise for Public Health," *The Journal of Law, Medicine & Ethics* 31:3 (Fall 2003): 379.

78 Ibid.: 379-80.

79 WHO, *Genomics and World Health*: 149.

80 For descriptions of case see U.S. Equal Employment Opportunity Commission (EEOC). Online: EEOC Homepage www.EEOC.gov/press/4-18-01.html.

81 Patricia A. Roche, "The Genetic Revolution at Work: Legislative Efforts to Protect Employees," *American Journal of Law & Medicine* 28 (2002): 276-77.

82 Ibid.: 276, with explanation by Dr. Chance in footnote 36.

83 "Analyzing Genetic Discrimination in the Workplace," *Human Genome News* 12:1, 2 (February 2002). Online: U.S. Department of Energy Office of Science Homepage www.ornl.gov/sci/techresources/Human_Genome/publicat/hgn/v12n1/09workplace.shtml.

84 U.S. Equal Employment Opportunity Commission (EEOC), "EEOC Petitions Court to Ban Genetic Testing of Railroad Workers in First EEOC Case Challenging Genetic Testing under Americans with Disabilities Act," press release (Feb. 9, 2001). Online: EEOC Homepage www.eeoc.gov.press.

85 Black, *War against the Weak*: 438.

86 Roche, "The Genetic Revolution at Work": 277.

## Chapter Three: A New Code to Crack

1 For statistics on Icelandic populations and descriptions of living conditions see Robert Kunzig, "Blood of the Vikings," *Discover* 19:12 (December 1998): 92.

2 Descriptions of Odin and Thor, ibid.

3 Gísli Pálsson, "The Life of Family Trees and the 'Book of Icelanders,'" *Medical Anthropology* 21:3-4 (July-December 2002): 337–67.

4 See Ingrid Wickelgren, *The Gene Masters: How a New Breed of Scientific Entrepreneurs Raced for the Biggest Prize in Biology* (New York: Times Books, 2002): 113.

5 Ibid.: 114.

6 Stefánsson citing Auden in Kunzig, "Blood of the Vikings": 97.

7 Stefánsson cited in "Gene Warrior," *New Scientist* 167:2247 (July 15, 2000): 42.
8 Description of Stefánsson's work in Chicago, Boston and Iceland, see Wickelgren, *The Gene Masters*: 115.
9 Skúlason cited in Pálsson, "The Life of Family Trees and the 'Book of Icelanders' ": 337–67.
10 Description of FRISK and DeCode project, see Gísli Pálsson, "Decoding Relatedness and Disease: The Icelandic Biogenetic Project," in *From Molecular Genetics to Genomics: The Mapping Cultures of Twentieth Century Genetics*, ed. Jean-Paul Gaudillière and Hans-Jörg Rheinberger (London: Routledge, 2004): 180–99.
11 For description of hunt for MS gene and DeCode start-up see Wickelgren, *The Gene Masters*: 114–22.
12 Thorgeirsson cited in Wickelgren, *The Gene Masters*: 114–22.
13 Gísli Pálsson and Paul Rabinow, "The Iceland Controversy: Reflections on the Trans-National Market of Civic Virtue," in *Global Assemblages: Politics and Ethics as Anthropological Problems*, ed. Aihwa Ong and Stephen J. Collier (Oxford: Blackwell, 2005): 91–103.
14 Hippocrates, "The Oath and Law of Hippocrates," *The Harvard Classics* XXXVIII:1 (New York: P.F. Collier & Son, 1909–1914). Online: Bartleby Homepage www.bartleby.com.
15 See Gísli Pálsson and Kristín E. Hardardóttir, "For Whom the Cell Tolls," *Current Anthropology* 43:2 (April 2002): 279.
16 Henry Greely, "Trusted Systems and Medical Records: Lowering Expectations," *Stanford Law Review* 52:5 (May 2000): 1585.
17 Jon F. Merz, Glenn E. McGee, and Pamela Sankar, " 'Iceland Inc.'?: On the Ethics of Commercial Population Genomics," *Social Science & Medicine*, 58:6 (2004): 1203.
18 For reaction to DeCode bill and Stefánsson's argument, see Wickelgren, *The Gene Masters*: 128–87.
19 Ibid.: 1203.
20 Interview with Halla Thorsteinsdottir; see also Pálsson and Hardardóttir, "For Whom the Cell Tolls": 278.
21 For details on the Iceland *Health Sector Database Act* see Merz, McGee and Sankar, " 'Iceland Inc.?' ": 1202.
22 Wickelgren, *The Gene Masters*: 180–84.
23 Merz, McGee and Sankar, " 'Iceland Inc.?' ": 1202–03.
24 DeCode press release, "DeCode Genetics Announces Year End 2002 Financial Results" (March 31, 2003). Online: DeCode Homepage www.decode.com.
25 DeCode press release, "DeCode Genetics Announces Third Quarter 2004 Financial Results" (November 2, 2004). Online: DeCode Homepage www.decode.com.
26 Stefánsson cited in John Greenwood, "Decoding Iceland's Genes," *National Post* (March 18, 2000): D1.
27 "Norse Code: A Fierce Debate in Iceland over Genetics Foreshadows Many Future Similar Battles Elsewhere," *Economist* 349:8097 (12/05/98): 99.

28 See Pálsson and Hardardóttir, "For Whom the Cell Tolls": 279.

29 UNESCO, *International Declaration on Human Genetic Data*, 32nd Plen. Sess., Gen. Conf., UNESCO Doc. C/29 Add.2 (2003).

30 Ibid.: 9.

31 Gísli Pálsson and Paul Rabinow, "The Icelandic Genome Debate," *Trends in Biotechnology* 19:5 (May 2001): 169.

32 Stefánsson cited in "Gene Warrior": 42.

33 For information on stroke gene and statistics see Nicholas Wade, "'Tour de Force' Finds Stroke Gene," *National Post*, reprinted from *The New York Times* (September 22, 2003): A1.

34 For description of osteoporosis gene see Sharon Begley, "Variant of Gene Increases Risk of Osteoporosis," *The Wall Street Journal* (November 3, 2003): B1.

35 Roger Boyes, "'Gene Bank' Set to Solve Riddle of Ill-Health," *London Times* (August 28, 2003): A19.

36 Arthur Caplan cited in Katharine Mieszkowski, "National Genes, Inc.," *Salon* (October 3, 2003). Online: Salon Homepage: www.Salon.com.

37 Ibid.

38 Phone interview with Bartha Knoppers.

39 UNESCO, Gen. Conf., 29th Plen. Sess., Res. 29, C/Res. 16, (1997), adopted by UN GA, G.A. res. 152, U.N. GAOR, 53rd Sess., U.N. Doc. A/RES/53/152 (1999).

40 *Council of Europe, C.A., Convention for the Protection of Human Rights and Dignity of the Human Being with regard to the Application of Biology and Medicine (Convention on Human Rights and Biomedicine)*, ETS No. 164 (1999). Online: Council of Europe Homepage www.conventions.coe.int/Treaty/en/Treaties/Html/164.htm.

41 Helen Frankish, "Coordination Centre for UK Biobank Project Announced," *The Lancet* 361:9370 (May 17, 2003): 1710.

42 See the UK Biobank Project website: UK Biobank Limited Homepage www.ukbiobank.ac.uk.

43 David E. Winickoff and Richard N. Winickoff, "The Charitable Trust as a Model for Genomic Biobanks," *The New England Journal of Medicine* 349:12 (September 18, 2003): 1180–84.

44 Eric Racine, "Discourse Ethics As an Ethics of Responsibility: Comparison and Evaluation of Citizen Involvement in Population Genomics," *Journal of Law, Medicine & Ethics* 31:3 (2003): 393–94.

45 Ricki Lewis, "Founder Populations Fuel Gene Discovery," *The Scientist* 15:8 (April 16, 2001). ) Online: The Scientist Homepage www.the-scientist.com.

46 Marc De Braekeleer and To-nga Dao, "In Search of Founders," *Human Biology* 66:2 (April 1994): 205–07.

47 Frederic Golden and Michael D. Lemonick, "The Race Is Over," *Time* 156:1 (July 3, 2000): 18–23.

48 Clinton cited in James D. Watson, *DNA: The Secret of Life* (New York: Alfred A Knopf, 2003): 191.
49 Watson cited in Michael D. Lemonick, "Gene Mapper," *Time* 156:26 (December 25, 2000/January 1, 2001): 110–15.
50 Watson, *DNA*: 193.
51 Bostein cited inWatson, *DNA*: 167–68.
52 Wyngaarden cited in Watson, *DNA*: 167–68.
53 Lemonick, "Gene Mapper": 110–15.
54 Leslie Roberts, "Controversial from the Start," *Science* 291:5507 (February 16, 2001): 1182.
55 Lemonick, "Gene Mapper": 110–15.
56 Collins cited in J. Madeleine Nash, "Riding the DNA Trail," *Time* 143:3 (January 17, 1994): 54–56.
57 For description of Francis Collins and Craig Venter see Wickelgren, *The Gene Masters*: 111.
58 Patton cited in Wickelgren, *The Gene Masters*: 111.
59 Collins cited in Golden and Lemonick, "The Race Is Over": 18–23.
60 Wickelgren, *The Gene Masters*: 134.
61 Venter cited in Wickelgren, *The Gene Masters*: 134.
62 For Perkin-Elmer Corp. deal and Celera launch, and Craig Venter's new deadline, see Wickelgren, *The Gene Masters*: 150–207.
63 For description of data released by Celera and HGP, see Roberts, "Controversial from the Start": 1182.
64 Clinton cited in Golden and Lemonick, "The Race Is Over": 18–23.
65 For description of Craig Venter and Francis Collins' agreement, and Celera's stock plunge, see Golden and Lemonick, "The Race Is Over": 18–23.
66 Ibid. See also Wickelgren, *The Gene Masters*: 267.
67 Carina Dennis, "The Rough Guide to the Genome," *Nature* 425:6960 (October 23, 2003): 758–59
68 Collins cited in Dennis, "The Rough Guide to the Genome": 758–59.
69 For description of sickle-cell discrimination see Jerry E. Bishop and Michael Waldholz, *Genome: The Story of the Most Astounding Scientific Adventure of Our Time—The Attempt to Map All the Genes in the Human Body* (New York: Simon & Schuster, 1990): 299.
70 National Institutes of Health (NIH) News Advisory, "Background on Ethical and Sampling Issues Raised by the International HapMap Project" (October 2002). Online: National Human Genome Research Institute Homepage www.genome.gov/10005337.
71 Phone interview with Bartha Knoppers.
72 For information on pharmacogenetics see Robert F. Mueller and Ian D. Young, *Emery's Elements of Medical Genetics* (London: Churchill Livingstone, 1998):

169–71; see also Nuffield Council on Bioethics, *Pharmacogenetics: Ethical Issues* (September 20, 2003): xiii–16

73 Nuffield Council on Bioethics, *Pharmacogenetics*: 16.

74 Organization for Economic Co-operation and Development (OECD), *Genetic Testing: Policy Issues for the New Millennium* (2000) 24. Online: OECD Homepage www.oecd.org.

75 Nuffield Council on Bioethics, *Pharmacogenetics*: 14–16. Online: Nuffield Council on Bioethics Homepage www.nuffieldbioethics.org.

76 Ibid.

77 Ibid.

## Chapter Four: Body Language

1 Quotations and descriptions of Michael Rudnicki's work from interviews with subject.

2 Haseltine cited in Nicholas Wade, *Life Script: How the Human Genome Discoveries Will Transform Medicine and Enhance Your Health* (New York: Touchstone, 2001): 119–20.

3 Quotation from interview with Pearl Campbell.

4 For background on EPO see Michael Sokolove, "The Lab Animal: Elite Athletes Always Have and Always Will Pursue Every Competitive Advantage—Health and the Law Be Damned. Is Genetic Manipulation Next?" *The New York Times Magazine* (January 18, 2004): 33.

5 Quotations and descriptions of Jane Aubin's work from interviews with subject.

6 Quotations and descriptions of Dr. Ernest McCulloch's work from interview with subject.

7 Quotation and description of James Till's life from interviews with James Till.

8 Phone interview with Max Cooper.

9 Ehrlich cited in Paul de Kruif, *Microbe Hunters* (San Diego: Harcourt Brace, 1996): 326.

10 Weissman cited in Peter Radetsky, "The Mother of All Blood Cells," *Discover* 16:3 (March 1995): 86–93.

11 For description of process see Steven Kotler, "The Final Frontier: Depending on Whom You Ask, Stem-Cell Research Is Either a Medical Godsend or Further Proof That God Is Dead," *LA Weekly* (January 31/February 6, 2003). Online: LA Weekly Homepage www.laweekly.com.

12 Radetsky, "The Mother of All Blood Cells": 86–93.

13 McCulloch cited in Radetsky, "The Mother of All Blood Cells": 86–93.

14 Interview with John D. Gearhart (June 18, 1999)). Online: Academy of Achievement Homepage www.achievement.org/autodoc/page/gea0int-1.

15 Ibid.

16 For description of James Thomson and his views on research on human embryos, see Frederic Golden, "Stem Winder: Before James Thomson Came Along, Embryonic Stem Cells Were a Researcher's Dream. Now They're a Political Hot

Potato." *Time* 158:7 (August 20, 2001). Online: Time Homepage www.time.com.

17 For description of culturing cells see Rick Weiss, "A Crucial Human Cell Isolated, Multiplied," *Washington Post* (November 6, 1998): A1.

18 Frost cited in "James Thomson: The Shy Pioneer of the Stem Cell," *People Weekly* 56:27 (December 31, 2001): 90.

19 Cohen cited in "Mending Hearts," *NBC Nightly News*, November 10 and 11, 2003.

20 Quotation and description of myoblast surgery from phone interview with Dr. Nabil Dib.

21 Dib cited in American Heart Association, "Muscle Cell Transplants Repair Damaged Heart Tissue," (November 17, 2002). Online: AHA Homepage www. american heart.org.

22 Robert Strumpf cited in Arizona Heart Institute, "To Care, to Teach, to Pioneer: Arizona Heart Institute Celebrates 30th Anniversary." Online: Arizona Heart Institute Homepage www.azheart.com.

23 Edward Diethrich cited in Arizona Heart Institute, "To Care, to Teach, to Pioneer."

## Chapter Five: The Body Builders

1 Shoichet cited in Industry Canada, *Follow the Leaders: Celebrating Canada's Biotechnology Innovators*, 2nd ed.: 15. Online: Invest in Canada Homepage investincanada.com.

2 For more information on the differences between human and mouse spinal cords see the Reeve-Irvine Research Center Homepage www.reeve.uci.edu/infodev.html.

3 For statistics on spinal cord injuries see, in Canada, the Canadian Paraplegic Association Homepage canparaplegic.org, and in the United States, The Spinal Cord Injury Resource Center Homepage www.spinalinjury.net.

4 Description of Molly Shoichet and quotations from Brian Shoichet from interview with Brian Shiochet.

5 Description of Shoichet's scaffolds from demonstration by former University of Toronto graduate student Kathryn Moore.

6 Vacanti cited in Catherine Arnst, "The Dynamic Duo of Tissue Engineering," *BusinessWeek* (July 27, 1998). Online: BusinessWeek Homepage www. businessweek.com.

7 Joseph D'Agnese, "Brothers with Heart: Four Brothers, Close since Childhood, Are Collaborating on the Development of Organs Made from a Patient's Own Tissue. Before Long, Custom-Grown Body Parts May Replace Transplants," *Discover* 22:7 (July 2001): 40; see also "The Mouse That Eared: Scientists Grow Human Ear on Back of Hairless Mouse," *The Gazette* (October 29, 1995): B4.

8 Anastasia Toufexis, "An Eary Tale," *Time* 146:19 (November 6, 1995): 60.

9 For information on the work and lives of the Vacanti brothers see D'Agnese, "Brothers with Heart," *Discover* 22:7 (July 2001): 38–102.

10 Jay Vacanti cited in D'Agnese, "Brothers with Heart": 39.
11 Robert Langer and Joseph Vacanti, "Tissue Engineering," *Science* 260:5110 (May 14, 1993): 920–26.
12 Monte Burke, "Plastic Man: MIT Kenneth J. Germeshausen Professor of Chemical and Biomedical Engineering Bob Langer Is Founder of Tissue Engineering," *Forbes* 170:13 (December 23, 2002): 296.
13 For description of Jay Vacanti and Robert Langer's work see D'Agnese, "Brothers with Heart": 39.
14 Jay Vacanti cited in D'Agnese, "Brothers with Heart": 39.
15 For description of Jay Vacanti's epiphany at Cape Cod see Dan Ferber, "Lab-Grown Organs Begin to Take Shape," *Science* 284:5413 (April 16, 1999): 422.
16 Jay Vacanti cited in Catherine Arnst and John Carey, "Biotech Bodies: Decades of Research into Tissue Engineering Are About to Pay Off As Dozens of Startups Perfect Living Organs Grown in the Lab, Not the Body," *Business Week* (July 27, 1998). Online: BusinessWeek Homepage www.businessweek.com.
17 For descriptions of McCormack's operation see Arnst and Carey, "Biotech Bodies."
18 McCormack cited in Arnst and Carey, "Biotech Bodies."
19 D'Agnese, "Brothers with Heart": 40.
20 Information on the LIFE Initiative from phone interview with Michael Sefton.
21 Quotations and opinions from interviews with John Davies.
22 For history of the restorative nature of bone see Robert F. Service, "Tissue Engineers Build New Bone," *Science* 289:5484 (September 1, 2000): 1498.
23 National Institute of Dental and Craniofacial Research (NIDCR), "Biomimetics and Tissue Engineering." Online: NIDCR Homepage www.nidcr.nih.gov/HealthInformation/OralHealthInformationIndex/SpectrumSeries/Biomimetics Tissue.htm.
24 S. Mendenhall, "The Bone-Graft Market in the United States," in *Bone Engineering*, ed. John Davies (Toronto: Em Squared, 2000): 585.
25 For information on scaffolds see Jeffrey M. Karp, Paul D. Dalton, and Molly S. Shoichet, "Scaffolds for Tissue Engineering," MRS Bulletin 28:4 (April 2003). Online: Materials Research Society Homepage www.mrs.org/publications/bulletin/2003/apr/.
26 Description of how to make scaffolds and grow bone cells in lab from demonstrations and interviews with Jeffrey Karp.
27 Ibid.
28 Robert S. Langer and Joseph P. Vacanti, "Tissue Engineering: The Challenges Ahead," *Scientific American* 280:4 (April 1999): 89.

# Notes

## Chapter Six: Heal Thyself

1 For information about different religious perspectives and the status of the embryo see Laura Shanner, "Stem Cell Terminology: Practical, Theological and Ethical Implications," *Health Law Review* 11:1 (December 2002): 62–65.
2 Lee M. Silver, *Remaking Eden: Cloning and Beyond in a Brave New World* (New York: Avon Books, 1997): 43.
3 For formulation of ethics see Ted Schrecker and Margaret A. Somerville, "Making Ethically Acceptable Policy Decisions: Challenges Facing the Federal Government," in *Renewal of the Canadian Biotechnology Strategy, Resource Document 3.4.1, Background Research Papers, Ethics* (Ottawa: Industry Canada, March 1998): 80.
4 Baier cited in Schrecker and Somerville, "Making Ethically Acceptable Policy Decisions": 80–87.
5 For definition of ethics and descriptions of ethical schools of thought see Michael W. Fox, *Bringing Life to Ethics: Bioethics for a Humane Society* (New York: State University of New York Press, 2001): 11.
6 Margaret Somerville, "Stud Bulls, Human Embryos, and the Politically Incorrect," *Queen's Quarterly* 108:1 (Spring 2001): 21–28.
7 Fox, *Bringing Life to Ethics*: 13-14.
8 Kevin E. Brown, *Genetic Engineering: Science and Ethics on the New Frontier* (New Jersey: Prentice Hall, 2001): 4-5.
9 T.L. Beauchamp, "Methods and Principles in Biomedical Ethics," *Journal of Medical Ethics* 29:5 (October 2003): 270–71.
10 Schrecker and Somerville, "Making Ethically Acceptable Policy Decisions": 73–74.
11 For opinions of Seymour Lipset see Schrecker and Somerville, "Making Ethically Acceptable Policy Decisions": 88.
12 For description of ethics of the marketplace and the American perspective see Somerville, "Stud Bulls, Human Embryos, and the Politically Incorrect": 21–28.
13 Phone interview with Françoise Baylis.
14 For some scenarios on cloning see Françoise Baylis, "Human Cloning: Three Mistakes and an Alternative," *Journal of Medicine and Philosophy* 27:3 (2002): 326.
15 Fox, *Bringing Life to Ethics*: 123.
16 Baylis, "Human Cloning": 320–21.
17 S.C. 2004, c. 2.
18 Sandel cited in William Fitzpatrick, "Surplus Embryos, Nonreproductive Cloning, and the Intend/Foresee Distinction," *The Hastings Center Report*, 33:3 (May/June 2003): 29–31.
19 Ibid.: 29–30.
20 The President's Council on Bioethics, *Reproduction and Responsibility: The Regulation of New Biotechnologies* (March 2004): 126 and 136–38. Online: The President's Council on Bioethics Homepage www.bioethics.gov/reports/reproduction andresponsibility/index.html.

21 1990 Chapter c. 37.
22 For U.K. Human Fertilisation and Embryology Act see Abhallah S. Daar and Lorraine Sheremeta, "The Science of Stem Cells: Some Implications for Law and Policy," *Health Law Review* 11:1 (December 2002): 8.
23 See Marie-Hélène Régnier and Bartha Maria Knoppers, "International Initiatives," *Health Law Review* 11:1 (December 2002): 67–71.
24 Ibid.: 70.
25 Ibid.
26 See Baylis, "Human Cloning": 323.
27 Bacon cited in Fox, *Bringing Life to Ethics*: 119–21.
28 For information on how governments can regulate science see Schrecker and Somerville, "Making Ethically Acceptable Policy Decisions": 84.
29 For opinions of Ronald Inglehart see Schrecker and Somerville, "Making Ethically Acceptable Policy Decisions": 89–90.
30 Ibid.: 83–84 and 99.
31 For legislative history of Canada's Assisted Human Reproduction Act, see Erik Parens and Lori P. Knowles, "Reprogenetics and Public Policy," *Hastings Center Report* (July/August 2003): s16–s17.
32 Stem Cell Network, "Senate Passes Bill C-6" (March 17, 2004). Online: Stemcell Network Homepage www.stemcellnetwork.ca.
33 Morin cited in Stem Cell Network, "Senate Passes Bill C-6."
34 Angela Campbell, "Defining a Policy Rationale for the Criminal Regulation of Reproductive Technologies," *Health Law Review* 11:1 (December 2002): 26–30.
35 Interviews with Timothy Caulfield.
36 Timothy Caulfield, "Bill C-13: The Assisted Human Reproduction Act: Examining the Arguments against a Regulatory Approach," *Health Law Review* 11:1 (December 2002): 21.
37 Yves Morin cited in Stem Cell Network, "Senate Passes Bill C-6."
38 Françoise Baylis *et al.*, "Cryopreserved Human Embryos in Canada and Their Availability for Research," *Journal of Obstetrics and Gynaecology Canada* 25:12 (December 2003): 1026–31.
39 G.W. Bush cited in President's Council on Bioethics, *Monitoring Stem Cell Research* (January 2004): 28. Online: The President's Council on Bioethics Homepage www.bioethics.gov/reports/stemcell/.
40 *Roe, et al. v Wade, District Attorney of Dallas County* 410 U.S. 113 (1973).
41 President's Council on Bioethics, *Monitoring Stem Cell Research*: 26.
42 Clinton cited in President's Council on Bioethics, *Monitoring Stem Cell Research*: 30.
43 G.W. Bush cited in President's Council on Bioethics, *Monitoring Stem Cell Research:* 32.
44 Michael D. Lemonick and John F. Dickerson, "Stem Cells in Limbo: Two Years after President Bush Said the U.S. Had All the Cell Lines It Needed, Where Did They Go?" *Time* 162:6 (August 11, 2003): 51.

45 Liza Dawson *et al.*, "Safety Issues in Cell-based Intervention Trials," *Fertility and Sterility* 80:5 (November 2003): 1077–84.

46 Ibid.

47 Ruth Faden cited in "Clinical Use of Embryonic Stem Cells Jeopardized by Bush's Funding," *Biotech Week* (December 3, 2003): 282.

48 President's Council on Bioethics, *Monitoring Stem Cell Research*: 37.

49 Figures based on the most recent survey available, using 2002 figures; see President's Council on Bioethics, *Monitoring Stem Cell Research*: 47.

50 Leonard Zon cited in Gareth Cook, "Stem Cell Center Eyed at Harvard: Researchers Seek to Bypass U.S. Restrictions," *The Boston Globe* (February 29, 2004): A1.

51 Alvin Powell, "From the Laboratory to the Patient: Stem Cell Institute Will Call on Expertise from around University to Turn Research into Therapy," *Harvard University Gazette* (April 22, 2004). Online: Harvard University Gazette Homepage www.news.harvard.edu/gazette.

52 Daley cited in Cook, "Stem Cell Center Eyed at Harvard": A1.

53 Dan Vergano, "Private Stem Cell Research Widens," *USA Today* (April 25, 2004). Online: USA Today Homepage www.usatoday.com.

54 Phone interview with Dr. Thomas Okarma.

## Chapter Seven: Golden Cells and Flying Pigs

1 Cynthia Robbins-Roth, *From Alchemy to IPO: The Business of Biotechnology* (Cambridge, Mass.: Perseus Publishing, 2000): 13–29.

2 Ibid.

3 For descriptions of Cohen, Boyer, and Swanson relationship see James D. Watson, *DNA: The Secret of Life* (New York: Alfred A. Knopf, 2003): 113–15.

4 For naming of Genetech see Watson, *DNA*: 114.

5 For business and growth of Genetech see Robbins-Roth, *From Alchemy to IPO*: 13–29; see also Watson, *DNA*: 113–15.

6 Robbins-Roth, *From Alchemy to IPO*: 18–19.

7 Gower cited in Robbins-Roth, *From Alchemy to IPO*: 18–19.

8 Ibid.: 20–22.

9 Watson, *DNA*: 125.

10 Robbins-Roth, *From Alchemy to IPO*: 186.

11 For description of turn in biotech industry see Robbins-Roth, *From Alchemy to IPO*: 186.

12 Solow cited in Michael J. Mandel, "The Health-Care Catastrophe That Won't Happen: Why Biotech Gains Will Rein in Medical Costs. The Lesson of Info Tech," *BusinessWeek,* 3885 (May 31, 2004): 57.

13 Cynthia L. Webb, "Beyond the Bubble," *The Washington Post* (April 7, 2003). Online: Washington Post Homepage www.washingtonpost.com.

14 Richard W. Oliver, *The Coming Biotech Age: The Business of Bio-Materials* (New York: McGraw-Hill, 2000): 159–61.

15 Herb Greenberg, "Placing a Bet on Biotech: Right Now This Risky Sector Is a Bust. That's Exactly Why I Think It's So Attractive," *Fortune* 146:13 (December 30, 2002): 188.

16 Gary Walsh, "Biopharmaceutical Benchmarks—2003," *Nature Biotechnology* 21:8 (August 2003): 865–70.

17 Chuck McNiven, Lara Raoub, and Namatié Traoré, *Features of Canadian Biotechnology Innovative Firms: Results from the Biotechnology Use and Development Survey, 2001* (report for Statistics Canada) (March 2003). Online: Statistics Canada Homepage www.statcan.ca.

18 Laura Common, "Advances in Biotech Will Be Breathtaking. So Will Costs," *Medicine 2010 Medical Post* (January 27, 2004) special supplement.

19 Oliver, *The Coming Biotech Age*: 157.

20 Sarah Staples, "DNA Testing by Tm Bioscience Is on the Cutting Edge of Medicine," *Canadian Business* 77:8 (April 12–25, 2004): 63.

21 Cutting Edge Information, *Pharmaceutical Alliances, Licensing and Deal-Making*: 82 and 99.

22 John T. Slania, "Payer, Beware; Biotech Drugs are the Latest Treatments, but They Can Cost Thousands per Dose, and They Don't Have Much of a Track Record. Employers and Insurers Ponder Where to Draw the Line," *Crain's Chicago Business* 26:33 (August 18, 2003): SR1.

23 Statistics and information on pentafuside, see Laura Common, "Advances in Biotech Will Be Breathtaking. So Will Costs."

24 McNiven, Raoub, and Traoré, *Features of Canadian Biotechnology Innovative Firms.*

25 For statistics on size of firms see Ernst & Young, *Resurgence: Global Biotechnology Report 2004—The Americas Perspective.* Online: Ernst & Young Homepage www.ey.com/global/content.nsf/International/Biotechnology_Reports_2004.

26 Bruce McConomy and Bixia Xu, "Value Creation in the Biotechnology Industry; As a Recent Study Demonstrates, It's Non-financial Disclosures That Are a Key to Value Creation in This and Other Growing Industries. Understand an Industry's Value Drivers and You'll Better Understand the Potential of Individual Firms," *CMA Management* 78:2 (April 2004): 28–30.

27 KPMG LLP, *The CEO's Guide to International Business Costs G-7 2004 Edition*, KPMG LLP (February 2004): 32.

28 Michael D. West, *The Immortal Cell: One Scientist's Quest to Solve the Mystery of Human Aging* (New York: Doubleday, 2003): 20–21.

29 Phone interview with Michael West.

30 Ibid.

31 West, *The Immortal Cell*: 49–53.

32 Ibid.: 61–62 and 88.

33 Ibid.

34 Ibid.: 89–99.

35 Ibid.

36 Ibid.: 92.

37 Rod McQueen, *The Last Best Hope: How to Start and Grow Your Own Business* (Toronto: McClelland & Stewart, 1995): 53–54.
38 McNiven, Raoub, and Traoré, *Features of Canadian Biotechnology Innovative Firms*: 18.
39 Ryan cited in West, *The Immortal Cell*: 94–98.
40 Phone interview with Miller Quarles.
41 Robbins-Roth, *From Alchemy to IPO*: 143.
42 Phone interview with West.
43 Ibid.
44 Oliver, *The Coming Biotech Age*: 150.
45 For requirements of a successful biotech see "Biotech after the IPO," *PR Week* (July 26, 2004): 15.
46 For pros and cons of going public see McQueen, *The Last Best Hope*: 260.
47 For information on disclosures see McConomy and Xu, "Value Creation in the Biotechnology Industry": 28–30.
48 For description of drug approval process see Raizel Robin, "Survival of the Fittest: the Economic Promise of Life Sciences Is Undeniable. But Some Major Hurdles Lie in the Way for Canadian Companies Competing in an International Field," *Canadian Business* 76:16 (September 1, 2003). Online: Canadian Business Homepage www.canadianbusiness.com.
49 Lambert cited in Scott Foster, "Biotechnology—Four Perspectives," *Ottawa Business Journal* (November 17, 2003). Online: Ottawa Business Journal Homepage www.ottawabusinessjournal.com.
50 Ernst & Young, *Resurgence*.
51 John DeLamarter, "Biotechnology Partnerships—Medicine for an Ailing Industry?" *Nature Biotechnology* 21:8 (August 2003): 847–48.
52 Robbins-Roth, *From Alchemy to IPO*: 166.
53 Statistics from SECOR, *Canadian Alliances in Biotechnology, April 2003*.
54 Andrew Morse, "Pharma Alliances: The Failure Rate," *The Deal* (January 9, 2004). Online: The Deal Homepage www.thedeal.com.
55 West, *The Immortal Cell*: 106–25.
56 For description of telomerase gene hunt see West, *The Immortal Cell*: 106–25.
57 For description of cloning and aging see West, *The Immortal Cell*: 166–69.
58 Phone interview with Thomas Okarma.
59 For information on GRN163 and GRN163L see "New Data Announced on the Effects of Telomerase Inhibitor Drugs," *Cancer Weekly* (January 6, 2004): 131.
60 For information on cancer vaccine see "Geron Reports Presentation of Further Results of Telomerase Cancer Vaccine Clinical Trial at Duke University Medical Center" (Geron press release). Online: Geron Homepage www.geron.com.
61 For statistics on drug development see Jorge Niosi, *Explaining Rapid Growth in Canadian Biotechnology Firms*, Statistics Canada (February 2000): 11.
62 Phone interview with Okarma.
63 West, *The Immortal Cell*: 177–79.

64 Jaenisch cited in Ed Welles, "Who Is Doctor West and Why Has He Got George Bush So Ticked Off," *Fortune* 145:9 (April 29, 2002): 104.
65 Shay cited in Welles, "Who Is Doctor West and Why Has He Got George Bush So Ticked Off": 104.
66 West cited in Ricki Lewis, "Mike West: Cloning for Human Therapeutics," *The Scientist* 16:18 (September 16, 2002): 60.
67 Phone interview with West.
68 Phone interview with West.

## Chapter Eight: Owning Ourselves (and Each Other)

1 Chakrabarty cited in Souvik Chowdhury, "Engineering Life," *The Hindu* (04/12/2002). Online: The Hindu Homepage hindu.com/thehindu/mp/2002/12/04/stories/2002120400310200.html.
2 Telephone interview with A.M. Chakrabarty.
3 A.M. Chakrabarty, "Patenting of Life-Forms: From a Concept to Reality," in *Who Owns Life?*, ed. David Magnus, Arthur Caplan, Glenn McGee (New York: Prometheus Books, 2002): 17–23.
4 A.M. Chakrabarty cited in Andrew Kimbrell, *The Human Body Shop: The Engineering and Marketing of Life* (New York: HarperSanFrancisco, 1993): 192–97.
5 Beth Burrows, "Patents, Ethics and Spin," in *Redesigning Life? The Worldwide Challenge to Genetic Engineering*, ed. Brian Tokar (New York: Zed Books Ltd., 2001): 239.
6 35 U.S.C. 100–376.
7 35 U.S.C. 101.
8 *Diamond, Commissioner of Patents and Trademarks v Chakrabarty*, 447 U.S. 303 (1980) at 308–09.
9 *Sinclair & Carroll Co., Inc. v Interchemical Corporation*, 325 U.S. 327 (1945) at 330–31.
10 Lorraine Sheremeta and Richard E. Gold, "Creating a Patent Clearinghouse in Canada: A Solution to Problems of Equity and Access," *Health Law Review* 11:3 (Fall 2003): 17.
11 *EC, Council Directive 98/44 of 6 July 1998 on the legal protection of biotechnological inventions* [1998] O.J. L. 213/13. Online: EU Homepage europa.eu.int.
12 European Patent Office, *Convention on the Grant of European Patents (European Patent Convention)*. Online: www.european-patent-office.org.
13 For information on European patents and restrictions see Stephen Wilkinson, *Bodies for Sale: Ethics and Exploitation in the Human Body Trade* (London: Routledge, 2003): 198.
14 For description of how patents fuel U.S. industry see Jack Wilson, "Patenting Organisms: Intellectual Property Law Meets Biology," in *Who Owns Life?*: 26–27.
15 For definition of a patent and criteria for patents see Sheremeta and Gold, "Creating a Patent Clearinghouse in Canada: A Solution to Problems of Equity and Access": 17.
16 Wilson, "Patenting Organisms": 27–28.

17 333 U.S. 127 (1948) at 130.
18 Wilkinson, *Bodies for Sale*: 194.
19 35 U.S.C. 161–64.
20 Luther Burbank cited in Wilson, "Patenting Organisms": 33–37.
21 Wilson, "Patenting Organisms": 35.
22 Ibid.: 28.
23 Kimbrell, *The Human Body Shop*: 193.
24 *Parker, Acting Commissioner of Patents and Trademarks v Flook*, 437 U.S. 584 (1978) at 596.
25 *Diamond, Commissioner of Patents and Trademarks v Chakrabarty.*
26 Ibid. at 316.
27 Ibid. at 307.
28 Ibid. at 308.
29 Ibid.; for definition of "manufacture" see *American Fruit Growers, Inc. v Brogdex Co.*, 283 U.S. 1 (1931) at 11; for definition of "composition of matter" see *Shell Development Co. v Watson*, 149 F. Supp. 279 (DC 1957) at 280.
30 *Diamond, Commissioner of Patents and Trademarks v Chakrabarty* at 309.
31 Ibid.
32 Ibid. at 310.
33 Chowdhury, "Engineering Life."
34 Jeremy Rifkin and Jeremy P. Tarcher, *The Biotech Century: Harnessing the Gene and Remaking the World* (New York: Putnam, 1998): 43.
35 For quotation and statistics on Genetech and Cetus see Kimbrell, *The Human Body Shop*: 196.
36 Chakrabarty cited in Chowdhury, "Engineering Life."
37 Donald Quigg cited in Wilson, "Patenting Organisms": 37.
38 Hatfield cited in Kimbrell, *The Human Body Shop*: 200.
39 Ibid.: 197.
40 R.S.C. 1985, c. P-4.
41 Canadian Biotechnology Advisory Committee, *Patenting of Higher Life Forms and Related Issues* (June 2002): 6. Online: CBAC Homepage cbac-cccb.ca.
42 *Harvard College v Canada (Commissioner of Patents)*, [2002] 4 S.C.R. 45 at para. 178.
43 *Harvard College v Canada (Commissioner of Patents)*, [2002] 4 S.C.R. 45 at para. 199.
44 Wayne Kondro, "Patenting Life. Canadian High Court Rejects OncoMouse," *Science* 298:5601 (December 13, 2002): 2112–13.
45 *Canadian Charter of Rights and Freedoms*, Part I of the *Constitution Act, 1982,* being Schedule B to the *Canada Act 1982* (U.K.), 1982, c. 11.
46 For aftermath of OncoMouse case in Canada see Janice Tibbetts, "Biotech Industry Warns Canada Behind World on Animal Patents," *National Post* (July 4, 2003): A5.
47 Statistics on breast cancer from the Canadian Cancer Society.
48 For descriptions of Mary-Claire King's work see James D. Watson, *DNA: The Secret of Life* (New York: Alfred A. Knopf, 2003): 313.
49 Canadian statistics for 2005 from the Canadian Cancer Society.

50 For statistics and descriptions of BRCA genes see Bryn Williams-Jones, "History of a Gene Patent: Tracing the Development and Application of Commercial BRCA Testing," *Health Law Journal* 10 (January 2002): 123–27.

51 King cited in Matthew Albright, *Profits Pending: How Life Patents Represent the Biggest Swindle of the 21st Century* (Maine: Common Courage Press, 2004): 12–15.

52 Ibid.: 14.

53 Williams-Jones, "History of a Gene Patent": 123–46.

54 Ibid.

55 Institut Curie, "Against Myriad Genetics's Monopoly on Tests for Predisposition to Breast and Ovarian Cancer Associated with the BRCA1 Gene" (September 26, 2002). Online: Institut Curie Homepage www.curie.fr/upload/presse/myriad opposition6sept01_gb.pdf. Descriptions of the inadequacies of Myriad's testing also provided by Dominique Stoppa-Lyonnet.

56 Institut Curie, "Against Myriad Genetics's Monopoly on Tests for Predisposition to Breast and Ovarian Cancer Associated with the BRCA1 Gene."

57 For description of impact of BRCA patents on Canada see Williams-Jones, "History of a Gene Patent": 123–46.

58 For Ontario reaction to BRCA patents, see editorial "The Patenting of Genes," *The Globe and Mail* (January 13, 2003): A12.

59 Arupa Ganguly cited in Albright, *Profits Pending*: 15.

60 Watson, *DNA*: 315.

61 Canadian College of Medical Geneticists, policy statement, "Patenting of the Human Genome" (September 21, 2002). Online: CCMG Homepage ccmg.medical.org.

62 Williams-Jones, "History of a Gene Patent": 123–46.

63 Institut Curie, "Against Myriad Genetics's Monopoly on Tests for Predisposition to Breast and Ovarian Cancer Associated with the BRCA1 Gene."

64 Dominique Stoppa-Lyonnet cited in Meredith Wadman, "Europe's Patent Rebellion," *Fortune* 144:6 (October 1, 2001): 44.

65 Kloiber cited in Albright, *Profits Pending*: 15.

66 For description of upstream and downstream process, see Albright, *Profits Pending*: 23.

67 Ibid.: 29.

68 Phone interview with Timothy Caulfield.

69 For eligibility of life patents see Nuffield Council on Bioethics, *The Ethics of Patenting DNA* (July 2002): 21–23. Online: Nuffield Council on Bioethics Homepage www.nuffieldbioethics.org.

70 Sandy M. Thomas, Michael M. Hopkins, and Max Brady, "Shares in the Human Genome—the Future of Patenting DNA," *Nature Biotechnology* 20:12 (December 2002): 1185–88.

71 UN GA Res. 217(III), UN GAOR, 3rd Sess., Supp. No. 13, U.N. Doc. A/810 (1948).

72 For statement and UN view on human rights see Canadian Biotechnology Advisory Committee, *Patenting of Higher Life Forms and Related Issues*: 8.

73 Mary Jane Mossman and William F. Flanagan, *Property Law: Cases and Commentary* (Toronto: Emond Montgomery Publications Ltd., 1998): 63.

74 Patricia Orwen, "The Profits of Gene Piracy," *Ottawa Citizen* (May 14, 1994): B4.
75 Kimbrell, *The Human Body Shop*: 207.
76 Wagner cited in Barbara J. Culliton, "Mo Cell Case Has Its First Court Hearing; The University of California Argues That a Patient Whose Tissues Were the Basis of a Patent Has No Valid Claim to a Share of Potential Profits," *Science* 226 (November 16, 1984): 813–14.
77 Gage cited in Culliton, "Mo Cell Case Has its First Court Hearing": 813.
78 Ibid.: 813–14.
79 Golde cited in Orwen, "The Profits of Gene Piracy": B4.
80 See Kimbrell, *The Human Body Shop*: 208.
81 Ibid.
82 *Moore v Regents of the University of California*, 793 P.2d 479 (Cal. 1990).
83 Ibid.: at 494–95 [footnotes omitted].
84 Moore cited in Orwen, "The Profits of Gene Piracy": B4.
85 Annas cited in Burrows, "Patents, Ethics and Spin": 245–46.
86 For idea of personal versus fungible property see Wilkinson, *Bodies for Sale*: 215–18.
87 Moore cited in Burrows, "Patents, Ethics and Spin": 246.
88 See especially Albright, *Profits Pending*: 29–30

## Chapter Nine: The Gene Thieves

1 See Joann C. Gutin, "End of the Rainbow," *Discover* 15:11 (November 1994): 70–71.
2 See Steve Olson, *Mapping Human History: Discovering the Past through Our Genes* (New York: Houghton Mifflin Company, 2002): 209.
3 Ibid.: 210–11.
4 Feldman and Dunston cited in Gutin, "End of the Rainbow": 70–74.
5 Harry cited in Stephen Wilkinson, *Bodies for Sale: Ethics and Exploitation in the Human Body Trade* (London: Routledge, 2003): 189.
6 Contreras cited in Patricia Kahn, "Genetic Diversity Project Tries Again," *Science* 266:5186 (November 4, 1994): 720–22.
7 Debra Harry, *The Human Genome Diversity Project and Its Implications for Indigenous Peoples* (briefing paper for Indigenous Peoples Council on Biocolonialism, January 1995. Online: IPCB Homepage www.ipcb.org/resolutions/htmls/pf2004.html.
8 For description of Sandimmun Neoral and Thermus aquaticus see Sarah Laird *et al.*, United Nations University, Institute of Advanced Studies, *Biodiversity Access and Benefit-sharing Policies for Protected Areas* (November 2003): 7. Online: IAS Homepage www.ias.unu.edu/binaries/UNUIAS_ProtectedAreasReport.pdf.
9 Commission on Intellectual Property Rights, *Integrating Intellectual Property Rights and Developing Policy* (London: Commission on Intellectual Property Rights, 2002): 76. Online: CIPR Homepage www.iprcommission.org/papers/pdfs/final_report/CIPRfullfinal.pdf.

10 For information on Da Vine patent see Commission on Intellectual Property Rights, *Integrating Intellectual Property Rights and Developing Policy*.
11 Rothschild cited in Matthew Albright, *Profits Pending: How Life Patents Represent the Biggest Swindle of the 21st Century* (Monroe, Maine: Common Courage Press, 2004): 45.
12 Vandana Shiva, *Biopiracy: The Plunder of Nature and Knowledge* (Toronto: Between the Lines, 1997): 73.
13 Acosta cited in Patricia Orwen, "The Profits of Gene Piracy," *The Ottawa Citizen* (May 14, 1994): B4.
14 Mooney cited in Orwen, "The Profits of Gene Piracy."
15 For description of Philippines' biodiversity see Daniel Alker *et al.*, *Governing Biodiversity: Access to Genetic Resources and Approaches to Obtaining Benefits from Their Use: The Case of the Philippines*. (Bonn: German Development Institute, 2002): 30–34. Online: Convention on Biological Diversity Homepage www.biodiv.org/doc/ case-studies.
16 "From Global Enclosure to Self Enclosure: Ten Years After—A Critique of the *CBD* and the 'Bonn Guidelines' on Access and Benefit Sharing," *Communique* 83 (January/February 2004). Online: ETC Group Homepage www.etcgroup.org.
17 Indigenous Peoples Council on Biocolonialism (IPCB), Third Session, UN Permanent Forum on Indigenous Issues, "Collective Statement of Indigenous Peoples on the Protection of Indigenous Knowledge" (New York, May 10–21, 2004). Online: IPCB Homepage www.ipcb.org/resolutions/htmls/pf2004.html.
18 For examples of biopiracy see Michael A. Gollin, "Biopiracy: The Legal Perspective," (February 2001). Online: www.actionbioscience.org. An updated version of Gollin's "Legal Consequences of Biopiracy," *Nature Biotechnology* 17 (September 1999).
19 See Dena S. Davis, "Genetics: The Not-so-new Thing," *Perspectives in Biology and Medicine* 47:3 (Summer 2004): 430–40.
20 For statistics on biological resources and patents see Davis, "Genetics: The Not-so-new Thing": 430–40.
21 Harry cited in Davis, "Genetics: the Not-so-new Thing": 430–40.
22 For description of Merck & Co. and INBio partnership see Alker *et al.*, *Governing Biodiversity*: 24; see also Victoria Tauli-Corpuz, "Biotechnology and Indigenous Peoples," in *Redesigning Life? The Worldwide Challenge to Genetic Engineering*, ed. Brian Tokar (New York: Zed Books, 2001): 274.
23 Vandana Shiva, "Biopiracy: The Theft of Knowledge and Resources": 285–88.
24 WWF International and Center for International Environmental Law (CIEL), *Biodiversity and Intellectual Property Rights: Reviewing Intellectual Property Rights in Light of the Objectives of the Convention on Biological Diversity* (March 2001): 3–5. Online: CIEL Homepage www.ciel.org/Publications/tripsmay01.PDF.
25 For description of Merck & Co. and INBio partnership see Alker *et al.*, *Governing Biodiversity*: 24–25.
26 Commission on Intellectual Property Rights, *Integrating Intellectual Property Rights and Developing Policy*.

27 WWF International and Center for International Environmental Law (CIEL), *Biodiversity and Intellectual Property Rights*: 3.

28 For description of CIPR see Albright, *Profits Pending*: 67.

29 For quotations and description of TRIPS see WWF International and Center for International Environmental Law (CIEL), *Biodiversity and Intellectual Property Rights*: 8–9.

30 World Health Organization and World Trade Organization, *WTO Agreements & Public Health: A Joint Study by the WHO and the WTO Secretariat* (2002): 87. Online: WTO Homepage www.wto.org.

31 For statistics and critical elements for essential medicines see World Health Organization and World Trade Organization, *WTO Agreements & Public Health*: 87–88.

32 See Caroline Thomas, "Trade Policy, the Politics of Access to Drugs and Global Governance for Health," in *Health Impacts of Globalization: Towards Global Governance*, ed. Kelley Lee (New York: Palgrave Macmillan, 2003): 181–85.

33 For description of parallel importing and compulsory licensing see Liani Kumaranayake and Sally Lake, "Regulation in the Context of Global Health Markets," in *Health Policy in a Globalising World*, ed. Kelley Lee, Kent Buse, and Suzanne Fustukian (Cambridge: Cambridge University Press, 2002): 86–87.

34 Margaret Duckett, *Compulsory Licensing and Parallel Importing: What Do They Mean? Will They Improve Access to Essential Drugs for People Living with HIV/AIDS?* (background paper for the International Council of AIDS Service Organizations, July 1999). Online: ICASO Homepage www.icaso.org/docs/compulsory english.htm.

35 Jeffrey Robinson, *Prescription Games: Money, Ego, and Power Inside the Global Pharmaceutical Industry* (Toronto: McClelland & Stewart, 2001): 37–38.

36 Love cited in Thomas, "Trade Policy, the Politics of Access to Drugs and Global Governance for Health": 182.

37 For statistics on ARVs see Kumaranayake and Lake, "Regulation in the Context of Global Health Markets": 87.

38 For AIDS statistics see Albright, *Profits Pending*: 68.

39 Ibid.: 69.

40 World Health Organization and World Trade Organization, *WTO Agreements & Public Health*: 106.

41 19 U.S.C. 2101.

42 For explanation of U.S. sanctions see Albright, *Profits Pending*: 69.

43 Garnier cited in Albright, *Profits Pending*: 70.

44 For explanation of South African case see World Health Organization and World Trade Organization, *WTO Agreements & Public Health*: 106.

45 Lewis cited in Albright, *Profits Pending*: 70.

46 Canadian HIV/AIDS Legal Network, "Victory in South Africa, but the Struggle Continues." Online: Canadian HIV/AIDS Legal Network Homepage www.aidslaw.ca.

47 Thomas, "Trade Policy, the Politics of Access to Drugs and Global Governance for Health": 186.

48 Nuffield Council on Bioethics, *The Ethics of Research Related to Healthcare in Developing Countries* (April 2002): 22. Online: Nuffield Council on Bioethics Homepage www.nuffieldbioethics.org.

49 Pan American Health Organization (PAHO), Program on Public Policy and Health, Division of Health and Human Development, *Trade in Health Services: Global, Regional, and Country Perspectives* (Washington: Pan American Health Organization, 2002): 52. Online: PAHO Homepage www.paho.org.

50 For statistics see World Health Organization (WHO) Advisory Committee on Health Research, *Genomics and World Health* (2002): 130. Online: WHO Homepage www.who.int/genomics.

51 Statistics ibid.: 124.

52 For examples of DNA technology used in developing countries see David J. Weatherall, "Genomics and Global Health: Time for a Reappraisal," *Science* 302 (October 24, 2003): 597–98.

53 Susan Ruttan, "Repairing Damaged Eyes: Stem Cell Research Gives New Hope to the Blind," CanWest News Service (June 15, 2005).

54 Elaine Carey, "Stem Cell Advance: Mount Sinai Hospital Scientists Bring the World Closer to Finding Cures for Such Diseases as Parkinson's and Diabetes," *Toronto Star* (June 9, 2005) A1.

# Glossary

**Adenine (A):** one of four DNA **bases**.

**Adeno-associated virus:** a parvovirus used as a **gene therapy vector**.

**Adenovirus:** a double-stranded **RNA virus** used as a **gene therapy vector**.

**Adult stem cell:** an undifferentiated cell that has the ability to copy itself, albeit to a limited degree, as well as differentiate into any type of specialized cell of the tissue in which it is located.

**Alleles:** different forms of the same **gene** found in a population. We inherit one allele of a gene from each parent; these can be identical or different alleles.

**Allograft:** a tissue graft from another person.

**Amino acid:** one of 20 different types of organic molecules that make up the structure of proteins.

**Antibody:** a **protein** created by the immune system to defend the body against an attacking **antigen**.

**Antigen:** a molecule that triggers **antibody** production.

**Apoptosis:** programmed death of a **cell**.

**Autograft:** a tissue graft taken from one area to treat an injury in another area of the same person.

**Axon:** the long fibre in a **neuron** (nerve cell) that transmits outgoing messages to neighbouring cells.

**Bacterium:** a unicellular organism; used in **biotechnology** to produce **proteins**.

**Base:** one of four substances (**adenine, cytosine, guanine,** and **thymine**) that make up **DNA**. The order of bases within DNA determines the structure of the **protein** produced.

**Biodiversity:** variety of species, both plant and animal.

**Bioinformatics:** the gathering, compiling, and analysis of biological information to create computer databases for study.

**Biopiracy:** appropriation of biological resources without the consent of the people whose resources are being taken, or without an offer to share benefits from such resources.

## Glossary

**Bioprospecting:** use of traditional knowledge and natural resources of the developing world for the benefit of the developed world.

**Biotechnology:** using **recombinant DNA,** cellular therapies and **gene therapies** to create commercial products, particularly medical therapies.

**Blastocyst:** an early **embryo,** composed of some 150 cells, not yet implanted in the wall of the uterus.

**Bone morphogenetic proteins (BMPs): proteins** that aid in fixing broken bones and encourage bone formation.

**BRCA1** and **BRCA2: genes** whose mutations can predispose the carrier to breast and ovarian cancers.

**Calcium phosphate:** the substance bone is composed of; calcium phosphate gives teeth and bone their hardness.

**CAT scan:** computerized axial tomography, an X-ray method that gives a cross-section and three-dimensional images of the body.

**Cell:** the basic unit of life, housing an organism's genetic information and other important material. A human has approximately 100 trillion cells, each with its own **nucleus** containing 23 pairs of **chromosomes,** containing the person's **DNA.**

**Cell-based therapies:** method that encourages **stem cells** to differentiate into a specific type of cell, providing a quantity of that cell type to repair damaged tissue.

**Cellular scaffold:** a biodegradable construct used in tissue engineering to encourage the body's natural regenerative capabilities to mend weakened and damaged body parts.

**Central nervous system (CNS):** the nerve tissues of the brain and spinal cord.

**Chromosome:** a threadlike structure where **genes** are found; the "instruction manual" that dictates a person's characteristics. Humans have 23 pairs of chromosomes, with one chromosome of each pair passed down from each parent.

**Chromosome disorder:** a condition that results from an excess or deficiency of genetic material. Down syndrome, for example, occurs because of an extra copy of a chromosome.

**Chromosome jumping:** an extension of **chromosome walking,** this is a technical trick that circularizes **DNA,** allowing scientists to focus on its end pieces and "jump" over the middle sections.

**Chromosome walking:** cloning overlapping pieces of **DNA** to isolate a vast region of DNA. The overlapping pieces are isolated from human DNA fragments contained in individual bacterial cells, which have the same **bases** (A, C, G, and T) as human DNA, and thus cannot distinguish it from its own DNA. The **bacteria** DNA replicates with these human bases.

**Clone:** (a) identical **DNA** pieces produced using **recombinant DNA** methods. (b) identical **cells** that originate from a single ancestor.

**Cloning:** a technique that creates identical copies of a **gene** or piece of **DNA,** which can be used to study genetic material.

**Collagen: protein** that gives skin its resilience and elasticity.

**Committed cell:** a cell whose development pathway has been determined, so that it can only differentiate into a particular type of cell.

**Complex-gene disorder:** disorder that is caused by the effects of several genes in combination with lifestyle factors. An example is heart disease.

**Cryopreservation:** storage of biological material at low temperatures so that it may be stored for a long time.

**Cytosine (C):** one of four **DNA bases**.

**Daughter cell:** offspring of a cell. A dividing **stem cell** produces daughter cells that can either continue dividing as empty vessels, or differentiate into a specialized cell with a particular function.

**Differentiation:** process of an unspecialized embryonic **cell** becoming a specialized type of cell, such as bone, blood, or muscle.

**DNA (deoxyribonucleic acid):** double-helix molecule found in the **nucleus** of most **cell** types, which passes an individual's characteristics on to descendants.

**DNA probe:** single-stranded DNA that is tagged (using, say, a radioactive label) so that it can be identified by researchers. Since a tagged single strand can be seen connecting to a complementary single strand of DNA, probes help identify **genes** within a DNA sample.

**Dominant gene disease:** disease caused by even one copy of a faulty gene, because the faulty gene "dominates" a normal gene it is paired with. *See* **Recessive gene disorder.**

**Dopamine:** a chemical messenger produced by nerve cells in the area of the brain called the substantia nigra. Its job is to relay messages from the substantia nigra to other parts of the brain, to control muscle movement.

**E. coli (*Escherichia coli*):** a **bacterium** used in genetics, which has a small **genome** and can be grown easily in a laboratory.

**Embryo:** in humans, the organism that exists from fertilization to the end of the eighth week of gestation, at which point it becomes a **fetus.**

**Embryonic germ cell:** a cell found in the gonadal area of the **embryo.**

**Embryonic stem cell:** unspecialized cell in the **embryo** that is capable of transforming into any one of many types of cells.

**Enzyme:** a type of **protein** that can speed up a chemical reaction.

**Erythropoietin (EPO):** a **hormone** manufactured by the kidneys that encourages bone marrow to form red blood cells.

**Eugenics:** the study and practice of selective breeding based on subjective and often arbitrary selection.

**Expressed sequence tag (EST):** a small piece of **DNA** that helps compose a complementary section of DNA, which is used to identify unknown genes and their location within the genome.

**Fetus:** the organism that develops from the **embryo** to the point of birth.

**Gene:** the functional and physical element of heredity that is spelled out using the **bases** A, C, T, and G. A gene regulates the formation of **proteins.**

**Gene bank:** a database of **DNA** used in **population genetics** to trace the origins of disease genes in the hopes of one day defeating many diseases.

**Gene mapping:** isolating **DNA** from blood or tissue samples and searching for unique patterns of DNA **bases** within different people who have a particular disease, to locate the disease gene.

**Gene therapy:** a technique that manipulates dysfunctional or non-functioning genes, or replaces them with healthy versions.

**Genetic disease:** a condition that results from a genetic mistake such as the mutation of a single **DNA base** or the disappearance or addition of an entire **chromosome.**

**Genetic engineering:** the process of isolating **DNA** from one **cell** and placing it in another cell, to modify the **genes** in an organism.

**Genetic marker:** a section of **DNA** whose location along a chromosome can be pinpointed and whose inheritance can be traced. It may or may not have a known genetic function. Markers are used to determine exact locations of **genes.**

**Genetic mutation:** an alteration of a **gene,** such as the flip of a single **DNA base** (a point mutation), or the disappearance of a base pair (deletion). Genetic mutations can result in dysfunctional gene expression, which in turn may cause non-functioning or dysfunctional **proteins.**

**Genetic test:** analysis of a person's **genes** to identify any genetic abnormalities or the addition or deletion of critical **proteins**. Such abnormalities can signal inherited risk for a particular genetic condition.

**Genome:** an organism's entire **DNA.** The human genome is compiled of 23 pairs of **chromosomes.**

**Genomics:** the examination of **genes** and their purpose, and the connection between genes, lifestyle factors, and **genetic disease.**

**Genotype:** an organism's genetic makeup, which is different than its **phenotype.**

**Guanine (G):** one of the four **DNA** bases.

**Haplotype:** the pattern of a person's alleles (forms of a gene) **single nucleotide polymorphisms** that is inherited as one unit.

**HapMap Project:** the creation of a **haplotype** map, to uncover the genetic variations between most individuals. Minuscule variations in **DNA** can influence people's risk for a particular disease.

**Hematopoietic stem cell:** type of **stem cell** that can transform into all types of white and red blood cells.

**Histology:** microscopic anatomy; the opposite of gross anatomy, the study of anatomy visible to the naked eye.

**Human Genome Diversity Project:** an international effort to understand diversity among humans by studying genetic variations within the species.

**Human Genome Project:** the sequencing and mapping of all the genes in the human genome.

**Immunodeficient:** having a deficient immune system due to an inherited genetic mistake, a disease, or the effects of an immunosuppressant drug.

**Insulin:** a hormone manufactured by the pancreas that is responsible for regulating the blood sugar level. Diabetes results when a person cannot generate enough insulin naturally.

**In vitro ("in glass"):** taking place in an artificial environment, usually a Petri dish or test tube.

**In vitro fertilization (IVF):** fertilization of an egg outside the female; *see* **In vitro.**

**Junk DNA:** sections of DNA whose sequences do not code for genes, but have other functions within the **genome.**

**Library:** a group of **clones** compiled from an organism's **genome.** A library might be made up of clones of every genetic sequence found in one particular **cell** type, or typical of a specific organ.

**Linkage:** the closeness of genetic markers along a **chromosome.** The closer these markers are, the greater is the likelihood that they will be inherited.

**Matrix:** the foundation that is required for new bone formation.

**Monoclonal antibody:** a purified type of antibody created by one single clone of cells.

**MRI scan:** magnetic resonance imaging, a radiology test that uses radio waves, computer technology, and magnetism to create images of the body.

**Myoblast:** a primitive type of muscle **cell** that aids in muscle repair.

**Neuron:** nerve **cell**; composed of a cell body, an **axon,** and dentrides.

**Nucleus:** part of a **cell** housing genetic information.

**Oct-4:** protein that targets genes that instruct cells to remain proliferating and undifferentiated for long periods of time.

**Oncogene:** a variety of **gene** associated with cancer, affecting **cell** growth.

**Ornithine transcarbamylase deficiency syndrome (OTC):** metabolic condition characterized by the liver's inability to rid the body of ammonia due to missing or malfunctioning enzymes within the liver.

**Osteoblast: cell** responsible for creating bone.

**Osteoclast: cell** responsible for breaking down bone.

**Osteoporosis:** loss of bone density that occurs when **osteoclast** activity exceeds **osteoblast** activity. This can be the result of a decrease in estrogen production after menopause.

**Parallel importing:** buying pharmaceuticals from a third party in another country, instead of from the manufacturer; to take advantage of lower prices. An exception to **TRIPS**.

**Parkinson's disease:** a motor system disorder that results in the loss of brain cells that are responsible for producing **dopamine**, whose job it is to control muscle movement. Parkinson's is characterized by trembling and rigidity of the body.

**Pathology:** the study of disease.

**PAX7:** a **protein** whose job it is to transform **adult stem cells** into specialized muscle cells during muscle regeneration. PAX7 is responsible for repairing damaged muscle tissue and could one day be used to treat degenerative diseases including muscular dystrophy.

**Peripheral nervous system (PNS):** nerve tissue outside the brain and spinal cord; the PNS joins the **central nervous system** to the rest of the body.

**Pharmacogenetics:** the science of creating custom-designed drug treatments. An individual's **genotype** is examined to determine effective drug therapies, using the knowledge that certain genotypes react differently to certain pharmaceuticals.

**Phenotype:** an organism's physical makeup. Phenotypes, or traits, include hair and eye colour.

**Plasmid:** a small molecule of **DNA** that contains **genes,** and is used to pass genes from one **bacterium** to another.

**Plasticity:** the ability of a **stem cell** from a certain tissue to transform into a cell from another tissue type.

**Pluripotency:** the ability of an individual **stem cell** to transform into any of several different types of cells. *See* **Totipotency**.

**Polymerase chain reaction (PCR):** a technique to detect a sequence of **bases** in **DNA.**

**Polymorphism:** a naturally occurring genetic variation that is harmless.

**Population genetics:** analysis of genetic variations among members of a certain population.

**Porosity:** in tissue engineering, the amount and type of open spaces or passageways within a structure such as a **cellular scaffold** or implant.

**Preimplantation genetic diagnosis (PGD):** genetic analysis of **embryos** during **in vitro fertilization,** before implantation and pregnancy, to avoid **genetic diseases** in at-risk families.

**Progenitor cell:** a primitive cell that can develop into many types of differentiated cells.

**Proliferation:** division of an individual **cell** into two identical **daughter cells**, enabling a cell population to grow.

**Protein:** molecule made by a **gene,** responsible for the form and function of **cells.** Examples include hormones, which regulate bodily function, and antibodies, which combat infection.

**Proteomics:** analysis of all the **proteins** encoded by an organism's **genes**.

**Recessive gene disease:** a disease that does not appear unless the person has a pair of faulty **genes,** because a normal gene will "dominate" the faulty gene. However, the faulty gene may be passed on to the next generation. *See* **Dominant gene disease.**

**Recombinant DNA: DNA** that is created by combining pieces of DNA from different organisms.

**Red blood cell:** type of blood cell that transports oxygen through the body.

**Regenerative medicine:** medicine assisting the body to repair itself, using its own **stem cells, genes,** and **proteins.** Used to replace old, injured, or weakened organs and other body parts.

**Reproductive cloning: cloning** for the purpose of creating a living organism.

**Restriction enzyme:** a type of **protein** that can identify certain **DNA** sequences, and cut DNA at these identified sites.

**Restriction fragment length polymorphism (RFLP):** the difference in **DNA** fragment sizes between individuals, with the DNA fragments cut by specific types of enzymes. Sequences are then used as markers for genetic **linkages.**

**Retrovirus:** a type of **virus** made of **RNA** rather than **DNA**. HIV is one example.

**Reverse transcriptase:** the process by which a **retrovirus** uses an enzyme to make a piece of **DNA** that is complementary to the **RNA** of the **retrovirus**; the double-stranded DNA is then added to a host cell.

**RNA (ribonucleic acid):** single-stranded molecule that works with **DNA** to compose a **cell**'s genetic material. Unlike DNA, which stores genetic data, RNA transports this data from the DNA to other areas within the cell, where the information is translated into a **protein.**

**Satellite cell:** type of cell located close to muscle fibres. If it becomes inactive, the muscle cannot grow.

**Self-renewal:** a technique used by **stem cells** to divide into non-specialized cells, multiplying the cell population.

**Sequencing:** identifying the order of **bases** in a piece of **DNA.**

**Sequence-tagged site (STS): DNA** sequence detected using polymerase chain reaction, used to create landmarks along the **genome.**

**Single-gene disorder:** condition that results from a mistake in one particular gene. Examples include cystic fibrosis and Tay-Sachs disease.

**Single nucleotide polymorphism (SNP):** variation of a **DNA** sequence when a nucleotide **base** (A, T, C, or G) within that sequence is changed.

**Somatic cell:** any of the body's cells except sperm or egg cells.

**Somatic cell nuclear transfer (SCNT):** removal of a nucleus from a **somatic cell,** and its insertion into an egg cell whose nucleus has already been removed. The **DNA** of the implanted nucleus controls the functions of the new cell. This was the process used to **clone** Dolly the sheep.

**Stem cell:** *see* **Adult stem cell, Embryonic stem cell.**

**Suppressor gene:** a **gene** that can suppress the functions of another gene.

**Telomerase:** an **enzyme** that controls the action of **telomeres,** activated in cancerous **cells** and inactivated in normal cells.

**Telomeres:** repeated **DNA** sequence found at the end of **chromosomes**, often described as resembling ends of shoelaces. Telomeres control the normal replication of chromosomes. When a **cell** divides, the telomere shortens. When there is no longer a telomere, the cell dies.

**Therapeutic cloning:** removal of cells from an individual, and addition of the cells to a fertilized enucleated egg cell. The egg is then left to divide, creating a **blastocyst. Stem cells** from the blastocyst are used in cellular therapies for diseases including **Parkinson's** and diabetes.

**Thymine (T):** one of four **DNA bases**.

**Tissue engineering:** a combination of genetic engineering and chemical engineering that constructs artificial tissue and organs. Tissue engineering builds new skin, bone, and cartilage for injured and elderly patients.

**Totipotency:** the ability of a **cell** to transform into any one of the numerous types of cells that compose a person or other multicellular organism. *See* **Pluripotency.**

**TRIPS (Trade-Related Aspects of Intellectual Property Rights):** an agreement drafted in 1994 by the World Trade Organization to provide the minimum standards for intellectual property protection for member countries of the WTO. The poorest members are granted exceptions to certain TRIPS rules.

**Undifferentiated cell:** a cell that has not yet become specialized. *See* **Differentiation.**

**Vector:** a vehicle used to transport **DNA** in **gene therapy** applications. Several types of **viruses,** including **retroviruses,** are used as vectors in humans.

**Virus:** a non-cellular organism, usually disease-carrying, that can only reproduce within a host **cell.**

**White blood cell:** a type of cell that combats infection from **bacteria** and **viruses**. There are many types of white blood cells.

**Whole genome shotgun approach:** method of mapping the genome in which random pieces are broken up and sequenced by a computer; the computer then rearranges the genome into its proper order.

**Wnt:** a protein secreted when tissue damage occurs, encouraging **stem cells** to divide and specialize into muscle cells.

Sources: National Institutes of Health, Industry Canada, Genome Canada, *Black's Medical Dictionary* 40th edition, ed. Gordon Macpherson (London: A&C Black, 2002).

# Bibliography

Albright, Matthew. *Profits Pending: How Life Patents Represent the Biggest Swindle of the 21st Century* (Monroe, Maine: Common Courage Press, 2004).

Alker, Daniel *et al.*, *Governing Biodiversity: Access to Genetic Resources and Approaches to Obtaining Benefits from Their Use: The Case of the Philippines.* (Bonn: German Development Institute, 2002). Online: Convention on Biological Diversity Homepage www.biodiv.org/doc/case-studies.

Anderson, W. French. "Human Gene Therapy: Scientific and Ethical Considerations." *Journal of Medicine & Philosophy* 10:3 (August 1985): 275.

Anderson, W. French. "The Best of Times, the Worst of Times." *Science* 288:5466 (April 28, 2000): 627.

Baylis, Françoise. "Betwixt and Between Human Stem Cell Guidelines and Legislation." *Health Law Review* 11:1 (December 2002): 44.

Baylis, Françoise *et al.* "Cryopreserved Human Embryos in Canada and Their Availability for Research." *Journal of Obstetrics and Gynaecology Canada* 25:12 (December 2003): 1026.

Bishop, Jerry E. and Michael Waldholz. *Genome: The Story of the Most Astounding Scientific Adventure of Our Time—The Attempt to Map All the Genes in the Human Body* (New York: Simon & Schuster, 1990).

Black, Edwin. *War against the Weak: Eugenics and America's Campaign to Create a Master Race* (New York: Four Walls Eight Windows, 2003).

Calvi, L.M. et al. "Osteoblastic Cells Regulate the Haematopoietic Stem Cell Niche." *Nature* 425 (October 23, 2003): 841.

Cameron, C. and R. Williamson. "Is There an Ethical Difference Between Preimplantation Genetic Diagnosis and Abortion?" *Journal of Medical Ethics* 29:2 (April 2003): 90.

Campbell, Angela. "Defining a Policy Rationale for the Criminal Regulation of Reproductive Technologies." *Health Law Review* 11:1 (December 2002): 26.

Caulfield, Timothy. "Bill C-13: The Assisted Human Reproduction Act: Examining the Arguments against a Regulatory Approach." *Health Law Review* 11:1 (December 2002): 20.

Commission on Intellectual Property Rights. *Integrating Intellectual Property Rights and Developing Policy* (London: Commission on Intellectual Property Rights, 2002). Online: CIPR Homepage www.iprcommission.org/papers/pdfs/final_report/CIPRfullfinal.pdf.

Daar, Abhallah S. and Lorraine Sheremeta. "The Science of Stem Cells: Some Implications for Law and Policy." *Health Law Review* 11:1 (December 2002): 5.

D'Agnese, Joseph. "Brothers with Heart: Four Brothers, Close since Childhood, Are Collaborating on the Development of Organs Made from a Patient's Own Tissue. Before Long, Custom-Grown Body Parts May Replace Transplants." *Discover* 22:7 (July 2001): 36.

Davies, John E. *Bone Engineering* (Toronto: Em Squared, 2000).

Dawson, Lisa, et al. "Safety Issues in Cell-Based Intervention Trials." *Fertility and Sterility* 80:5 (November 2003): 1077.

de Kruif, Paul. *Microbe Hunters* (San Diego: Harcourt Brace, 1996).

Dennis, Carina. "The Rough Guide to the Genome." *Nature* 425:6960 (October 23, 2003): 758.

Department of Health and Human Services, National Institutes of Health, Recombinant DNA Advisory Committee, Minutes of Meeting, June 8–9, 1995. Online: NIH Homepage www4.od.nih.gov/oba/rac/minutes/6-8-9-95.htm.

Ernst & Young, *Resurgence: Global Biotechnology Report 2004—The Americas Perspective.* Online: Ernst & Young Homepage www.ey.com/global/content.nsf/International/Biotechnology_Reports_2004.

Florencio, Patrik S. and Mark K. Searl. "Genetic Interventions in Minors and Designer Babies." *Health Law in Canada* 21:2 (November 2000): 51.

Florencio, Patrik S. and Erik D. Ramanathan. "Secret Code: The Need for Enhanced Privacy Protections in the United States and Canada to Prevent Employment Discrimination Based on Genetic and Health Information." *Osgoode Hall Law Journal* 39:1 (2001): 77.

Fox, Michael W. *Bringing Life to Ethics: Bioethics for a Humane Society* (New York: State University of New York Press, 2001).

Golden, Frederic and Michael D. Lemonick. "The Race Is Over." *Time* 156:1 (July 3, 2000): 18.

Goldstein, David B. et al. "Pharmacogenetics Goes Genomic." *Nature Genetics* 4 (December 2003): 937.

Grace, Eric S. *Biotechnology Unzipped: Promises and Realities.* (Toronto: Trifolium Books Inc., 1997).

Harper, Peter. *Practical Genetic Counselling* (Oxford: Butterworth and Heineman, 1998).

Hubbard, Ruth and Eliiah Wald. *Exploding the Gene Myth: How Genetic Information Is Produced and Manipulated by Scientists, Physicians, Employers, Insurance Companies, Educators, and Law Enforcers* (Boston: Beacon Press, 1999).

International Conference on Harmonisation of Technical Requirements for Registration of Pharmaceuticals for Human Use. "ICH Harmonised Tripartite Guideline:

Guideline for Good Clinical Practice." (May 1, 1996): Online: ICH Homepage www.ich.org.

Jimenez Hamann, Maria C. et al. "Novel Intrathecal Delivery System for Treatment of Spinal Cord Injury." *Experimental Neurology* 182 (2003): 300.

Jones, Steve. *In the Blood: God, Genes and Destiny* (New York: HarperCollins, 1996).

Kimbrell, Andrew. *The Human Body Shop: The Engineering and Marketing of Life* (New York: HarperSanFrancisco, 1993).

Kunzig, Robert. "Blood of the Vikings." *Discover* 19:12 (December 1998): 90.

Laird, Sarah *et al.* United Nations University, Institute of Advanced Studies, *Biodiversity Access and Benefit-Sharing Policies for Protected Areas* (November 2003). Online: IAS Homepage www.ias.unu.edu/binaries/UNUIAS_ProtectedAreasReport.pdf.

Langer, Robert and Joseph Vacanti. "Tissue Engineering." *Science* 260:5110 (May 14, 1993): 920.

Langer, Robert S. and Joseph P. Vacanti. "Tissue Engineering: The Challenges Ahead." *Scientific American* 280:4 (April 1999): 89.

Lemonick, Michael D. "Gene Mapper." *Time* 156:26 (December 25, 2000/January 1, 2001): 110.

Lippman, Abby. "Led (Astray) by Genetic Maps: The Cartography of the Human Genome and Health Care." *Social Science and Medicine* 35:12 (1992): 1469.

Lippman, Abby. "Prenatal Genetic Testing and Screening: Constructing Needs and Reinforcing Inequities." *American Journal of Law & Medicine.* 17:1, 2 (1991): 15.

Lyon, Jeff and Peter Gorner. *Altered Fates: Gene Therapy and the Retooling of Human Life.* (New York: W. W. Norton & Co., 1995).

Maddox, Brenda. *The Dark Lady of DNA* (New York: Perennial, 2002).

Magnus, David, Arthur Caplan, Glenn McGee, ed. *Who Owns Life?* (New York: Prometheus Books, 2002).

Marshall, Eliot. "Clinical Trials: Gene Therapy Death Prompts Review of Adenovirus Vector." *Science* 286:5448 (December 17, 1999): 2244.

McNiven, Chuck, Lara Raoub, and Namatié Traoré. *Features of Canadian Biotechnology Innovative Firms: Results from the Biotechnology Use and Development Survey, 2001* (report for Statistics Canada) (March 2003). Online: Statistics Canada Homepage www.statcan.ca.

Medical Research Council of Canada, Natural Sciences and Engineering Research Council of Canada, Social Sciences and Humanities Research Council of Canada. "Tri-Council Policy Statement: Ethical Conduct for Research Involving Humans." (1998, with 2000 and 2002 updates): 4.1. Online: Interagency Advisory Panel On Research Ethics Homepage www.pre.ethics.gc.ca.

Morange, Michel *The Misunderstood Gene* (Cambridge: Harvard University Press, 2002).

Mueller, Robert F. and Ian D. Young. *Emery's Elements of Medical Genetics.* 11th ed. (London: Churchill Livingstone, 2001).

Nuffield Council on Bioethics. *The Ethics of Research Related to Healthcare in Developing Countries* (April 2002). Online: Nuffield Council on Bioethics Homepage www.nuffieldbioethics.org.

Nuffield Council on Bioethics. *Genetic Screening: Ethical Issues* (December 1993). Online: Nuffield Council on Bioethics Homepage www.nuffieldbioethics.org.

Nuffield Council on Bioethics. *The Ethics of Patenting DNA* (July 2002). Online: Nuffield Council on Bioethics Homepage www.nuffieldbioethics.org.

Oliver, Richard W. *The Coming Biotech Age: The Business of Bio-Materials* (New York: McGraw-Hill, 2000).

Olson, Steve. *Mapping Human History: Discovering the Past through Our Genes* (New York: Houghton Mifflin Company, 2002).

Ontario Provincial Advisory Committee on New Predictive Genetic Technologies. *Genetic Services in Ontario: Mapping the Future* (Nov. 30, 2001).

Online: Ontario Ministry of Health and Long-Term Care Homepage www.health.gov.on.ca.

Organization for Economic Co-operation and Development (OECD). *Genetic Testing: Policy Issues for the New Millennium* (2000). Online: OECD Homepage www.oecd.org.

Pálsson, Gísli. "The Life of Family Trees and the 'Book of Icelanders.'" *Medical Anthropology* 21:3-4 (July-December 2002): 337.

Pálsson, Gísli. "Decoding Relatedness and Disease: The Icelandic Biogenetic Project." in *From Molecular Genetics to Genomics: The Mapping Cultures of Twentieth Century Genetics* ed. Jean-Paul Gaudillière and Hans-Jörg Rheinberger (London: Routledge, 2004): 180.

Pálsson, Gísli and Paul Rabinow. "The Iceland Controversy: Reflections on the Trans-National Market of Civic Virtue." in *Global Assemblages: Politics and Ethics as Anthropological Problems* ed. Aihwa Ong and Stephen J. Collier (Oxford: Blackwell, 2005): 91.

Pan American Health Organization (PAHO), Program on Public Policy and Health, Division of Health and Human Development. *Trade in Health Services: Global, Regional, and Country Perspectives.* (Washington: Pan American Health Organization, 2002). Online: PAHO Homepage www.paho.org.

President's Council on Bioethics. *Reproduction and Responsibility: The Regulation of New Biotechnologies* (March 2004). Online: The President's Council on Bioethics Homepage www.bioethics.gov/reports/reproductionandresponsibility/index.html.

President's Council on Bioethics. *Monitoring Stem Cell Research* (January 2004). Online: The President's Council on Bioethics Homepage www.bioethics.gov/reports/stemcell/.

Radetsky, Peter. "The Mother of All Blood Cells." *Discover* 16:3 (March 1995): 86.

Régnier, Marie-Hélène and Bartha Maria Knoppers. "International Initiatives." *Health Law Review* 11:1 (December 2002): 67.

Revkin, Andrew. "Hunting Down Huntington's: From the Shores of Maracaibo to the Halls of Washington, Nancy Wexler Has Spent 25 Years Stalking Her Mother's Killer." *Discover* 14:12 (December 1993) 98.

Ridley, Matt. *Genome: The Autobiography of a Species in 23 Chapters.* (New York: HarperCollins, 1999).

Ridley, Matt. *Nature via Nurture: Genes, Experience & What Makes us Human* (Toronto: HarperCollins, 2003).

Rifkin, Jeremy and Jeremy P. Tarcher. *The Biotech Century: Harnessing the Gene and Remaking the World* (New York: Putnam, 1998).

Robbins-Roth, Cynthia. *From Alchemy to IPO: The Business of Biotechnology* (Cambridge, Mass.: Perseus Publishing, 2000).

Robinson, Jeffrey. *Prescription Games: Money, Ego, and Power Inside the Global Pharmaceutical Industry* (Toronto: McClelland & Stewart, 2001).

Roche, Patricia A. "The Genetic Revolution at Work: Legislative Efforts to Protect Employees." *American Journal of Law & Medicine* 28 (2002) 271.

Schrecker, Ted and Margaret A. Somerville. "Making Ethically Acceptable Policy Decisions: Challenges Facing the Federal Government." in *Renewal of the Canadian Biotechnology Strategy, Resource Document 3.4.1, Background Research Papers, Ethics* (Ottawa: Industry Canada, March 1998).

Rzeszutek, K. et al. "Proton Pump Inhibitors Control Osteoclastic Resorption of Calcium Phosphate Implants and Stimulate Increased Local Reparative Bone Growth." *The Journal of Craniofacial Surgery* 14:3 (May 2003):301.

Service, Robert F. "Tissue Engineers Build New Bone." *Science* 289:5484 (September 1, 2000): 1498.

Shanner, Laura. "Stem Cell Terminology: Practical, Theological and Ethical Implications," *Health Law Review* 11:1 (December 2002): 62.

Shiva, Vandana. *Biopiracy: The Plunder of Nature and Knowledge* (Toronto: Between the Lines, 1997).

Silver, Lee M. *Remaking Eden: Cloning and Beyond in a Brave New World* (New York: Avon Books, 1997).

Silvers, Anita and Michael Ashley Stein. "Human Rights and Genetic Discrimination: Protecting Genomics' Promise for Public Health." *The Journal of Law, Medicine & Ethics* 31:3 (Fall 2003): 377.

Somerville, Margaret. "Stud Bulls, Human Embryos, and the Politically Incorrect." *Queen's Quarterly* 108:1 (Spring 2001): 21.

Stolberg, Sheryl Gay. "The Biotech Death of Jesse Gelsinger." *New York Times Magazine* (November 28, 1999): 136.

Tokar, Brian, ed. *Redesigning Life? The Worldwide Challenge to Genetic Engineering* (New York: Zed Books Ltd., 2001).

Watson, James D. *DNA: The Secret of Life*. (New York: Alfred A Knopf, 2003).

Wade, Nicholas. *Life Script: How the Human Genome Discoveries Will Transform Medicine and Enhance Your Health* (New York: Touchstone, 2001).

Weatherall, David J. "Genomics and Global Health: Time for a Reappraisal." *Science* 302 (October 24, 2003): 597.

West, Michael D. *The Immortal Cell: One Scientist's Quest to Solve the Mystery of Human Aging* (New York: Doubleday, 2003).

Wexler, Alice. *Mapping Fate: A Memoir of Family, Risk, and Genetic Research.* (New York: University of California Press, 1996).

Wexler, Nancy S. "Will the Circle Be Unbroken? Sterilizing the Genetically Impaired." in *Genetics and the Law II* ed. Aubrey Milunsky and George J. Annas (New York: Plenum Press, 1979): 313.

Wexler, Nancy S. "Clairvoyance and Caution: Repercussions from the Human Genome Project." in *Code of Codes: Scientific and Social Issues in the Human Genome Project* ed. Daniel J. Kevles and Leroy Hood (Cambridge: Harvard University Press, 1992): 211.

Wickelgren, Ingrid. *The Gene Masters: How a New Breed of Scientific Entrepreneurs Raced for the Biggest Prize in Biology* (New York: Times Books, 2002).

Wilkinson, Stephen. *Bodies for Sale: Ethics and Exploitation in the Human Body Trade* (London: Routledge, 2003).

Williams-Jones, Bryn. "History of a Gene Patent: Tracing the Development and Application of Commercial BRCA Testing." *Health Law Journal* 10 (January 2002): 123.

Winickoff, David E. and Richard N. Winickoff. "The Charitable Trust as a Model for Genomic Biobanks." *The New England Journal of Medicine* 349:12 (September 18, 2003): 1180.

World Health Organization (WHO) Advisory Committee on Health Research, *Genomics and World Health* (2002). Online: WHO Homepage www.who.int/genomics.

WWF International and Center for International Environmental Law (CIEL). *Biodiversity and Intellectual Property Rights: Reviewing Intellectual Property Rights in Light of the Objectives of the Convention on Biological Diversity* (March 2001). Online: CIEL Homepage www.ciel.org/Publications/tripsmay01.PDF.

# Index


abortion, 48–50, 158, 160. *See also* embryos
Acosta, Isidro, 230
Action Group on Erosion, Technology, and Concentration. *See* ETC Group
adeno-associated virus (AAV), 8–9
adenosine deaminase (ADA), 27
adenovirus, 7, 20
Advanced Cell Technology (ACT), 149, 190, 194–95
Affymetrix, 86–87, 91
Africa, 241–42
AIDS, 46, 93, 174, 239–41, 242, 243. *See also* HIV
Albright, Matthew, 216
allografts, 135–36
Alzheimer's disease, 244–45
American Breeders Association, 40
American College of Medical Geneticists, 214
American Eugenics Society, 42
Amgen, 170, 187
amniocentesis, 46
Amoxil, 238
amyotrophic lateral sclerosis (ALS, or Lou Gehrig's disease), 62
Anderson, W. French, 26–27
anemia, 92
  sickle-cell, 53, 83
Annas, George, 221
antibodies, 105–6
antigens, 192
anti-retroviral (ARV) therapies, 239–41
Arizona Heart Institute, 110–15
Arkwright, Richard, 197
Arley, Niels, 102–3
Assisted Human Reproduction Agency of Canada (AHRAC), 159
Association for an Upright Generation, 52
Aubin, Jane, 93–97
Austria, 152
autografts, 120, 135
autonomy, 50, 51, 146
Avery, Oswald, 32
ayahuasca, 228–29

Bacon, Francis, 153–54
bacteria, 168, 228
  patents on, 196–97, 199–204
Baier, Annette, 144
*Banisteriopsis*, 228–29
Barshefsky, Charlene, 241
basmati rice, 232
Bastarache, Michel, 207–8
Bates, Gillian, 38
Batshaw, Mark, 20, 21
Baylis, Françoise, 147, 149, 153, 159
Beauchamp, T.L., 146
Bell, Eugene, 128
Benedickz, John, 63
Bessis, Roger, 50
Binnie, Ian, 208
biocolonialism. *See* biopiracy
biodiversity, 230–32, 234–35
bioethics, 147. *See also* ethics
Biogen Inc., 28
bioinformatics, 91
biomaterials, 120
biopharmaceuticals, 173–74
biopiracy, 217–22, 224–45
bioprospecting, 68–69, 227–32, 233–35
biotech industry, 166–95
  accountability of, 183–84
  in Canada, 173, 174, 180, 184–85
  investment in, 170–72, 180–85, 193, 194, 198, 204
  and patents, 198, 212–15
  and pharmaceutical industry, 184–86
  products, 172–73
  R&D in, 175, 185, 198
  revenues, 174, 184
Black, Edwin, 42
Blair, Tony, 76, 80
blastocysts, 88–89, 148
blindness, 244
bone, 132–40
bone marrow, 100–101, 105, 132, 139
BoneTech, 137
*Book of Icelanders,* 63
Botstein, David, 77
Boyer, Herb, 167–70
BRACAnalysis, 211–15
brain cancer, 128
brain imaging, 13
BRCA genes, 46, 75, 208–15. *See also* breast cancer
  patents on, 210–11, 213
  screening for, 211–15
breast cancer, 208–15. *See also* BRCA genes
  BRCA1 gene and, 46, 75, 208–11
  BRCA2 gene and, 211–15
  HER2 and, 84–85
  occurrence, 53, 209
Breast Cancer Linkage Consortium (BCLC), 210

# Index


British Columbia, 213
Brock, Thomas, 228
Brown, David, 87
Buchwald, Manuel, 43–44
Burbank, Luther, 200
Burger, Warren, 202–3
Burlington Northern Santa Fe Railroad Co., 57–58
Bush, George W., 151, 161–62, 163–64, 184

California, 152, 219–20
Cameron, C., 49
Campbell, Angela, 156–57
Campbell, Pearl, 91–92
Canada. *See also* Canada: legislation
  biotech industry in, 173, 174, 180, 184–85
  cancers in, 209
  cloning regulation in, 149–51, 155–59
  ethics in, 146
  eugenics in, 41
  gene therapy in, 23
  genetic testing in, 212–14
  heart disease in, 114
  and International HapMap Project, 82–83
  Networks of Centres of Excellence, 89
  and OncoMouse, 206
  patent legislation in, 198, 206–7
  patent office, 206–7
  population genetics in, 74–75
  research in, 24, 26, 82–83
  spinal cord injury in, 119–20
  stem cell research in, 155–59, 244
Canada: legislation
  Assisted Human Reproduction Act (2004), 149–51, 156–59
  Charter of Rights and Freedoms, 51, 157, 208
  Patent Act, 206–7
  Personal Information Protection and Electronic Documents Act (PIPEDA), 55
Canadian Biotechnology Advisory Committee, 217
Canadian College of Medical Geneticists, 214
cancer, 33–34, 105, 186–87, 209, 210, 217. *See also specific types of cancer*
Cantor, Charles, 224
Caplan, Arnold I., 140
Caplan, Arthur, 17, 72
Capron, Alexander, 24
cardiac disease. *See* heart disease
carpal tunnel syndrome, 57–58
CARTaGENE, 74
cartilage, 125–26
Caulfield, Timothy, 157–58, 163–64, 216
Cavalli-Sforza, Luca, 224, 225
Celera Genomics, 79, 216
cell lines, 141, 164–65, 244
  available, 159, 161–62
  corporate control of, 162–64
  patents on, 217–22, 230
cells. *See also specific types of cells and components;* cell lines; cloning
  aging of, 176–82
  animal, 140
  bone marrow, 100–101, 105
  cancer, 186–87
  creation of, 128
  germ (reproductive), 107–8
  mesenchymal, 140
  muscle, 134
  nerve, 123–24, 244
  osteogenic, 139–40
  production of, 140–42
  progenitor, 92–93, 134, 139–40, 192
  proteins on, 105–6
  radiation and, 100, 101, 102–3
  regeneration of, 191
  reproduction of, 102–4, 108–9
  skin, 128
  somatic, 148
  therapies using, 188–90
  transplantation of, 105–6, 110–13, 141–42
Center for International Environmental Law, 229
Centre for Stem Cell and Gene Therapy, 86–87, 91
Cetus Corp., 167, 169, 204
Chakrabarty, Ananda Mohan, 196–97, 199–204
Challenges in Regenerative Medicine (CHARM), 1
Chance, Phillip, 58
Chapman, Karen, 189, 195
Charo, R. Alta, 108
chromosomes, 32, 44–45, 64. *See also* DNA; telomeres
Cibelli, Jose, 190
Cleveland Clinic, 140
clinical trials, 21–23, 171, 193
  of drugs, 184, 192, 193
  for gene therapy, 27
  guidelines for, 21–22, 23
  of myoblast transplantation, 110–13
  of OTC gene therapy, 17–21, 22, 28
  of Parkinson's gene therapy, 11–16
  reporting requirements of, 23–24, 30
Clinton administration, 161, 240, 241
Clinton, Bill, 57, 76, 80, 160, 161, 181
clones, 44, 101
cloning, 148–49, 189–90
  human, 149, 155, 194–95
  regulation of, 149–53, 155–59
  reproductive, 148, 158
  therapeutic, 148, 149, 158–59, 164
Cohen, Samuel, 110–13, 115
Cohen, Stanley, 167–70
"Collective Statement of Indigenous Peoples on the Protection of Indigenous Knowledge," 232
Collins, Francis, 44–45, 76, 78–81, 82, 210
colony-forming units (CFUs). *See* stem cells
Commission on Intellectual Property Rights (CIPR), 236
Concepción, Maria, 36
conflict of interest, 25–26, 28–29
consent (informed)
  in clinical trials, 20–21, 22–23
  for genetic testing, 57–58
  medical data and, 65–66, 67, 69, 71–72, 73–74
  in Moore cell line case, 218–19
Conservation International, 231
Contreras, Rodrigo, 226
Convention on Biological Diversity (CBD), 234–35
Cook, Howard, 179
Cooper, Max, 104–5
Coordinating Body of Indigenous Organizations of the Amazon Basin (COICA), 229
Costa Rica National Biodiversity Institute (INBio), 233–34

# Index


Crick, Francis, 32, 33, 76
Curing Old Age Disease Society, 180
cyclosporine, 227
cystic fibrosis (CF), 34, 42–45, 46, 48, 62

Daley, George Q., 164
Darwin, Charles, 232
databases, 91, 210. *See also specific databases*
  access to, 65–67, 69, 72–73, 214–15
  genealogical, 63–64, 65–69
  on gene therapy, 24, 30
David Oddsson, 63–64
Davies, John "Jed," 131–34, 135, 136, 137, 141
DaVine plant, 228–29
De Bakey, Michael, 115
DeCode Genetics, 64–65, 67–68, 70
de Kruif, Paul, 105
deontology, 145–46
de Silva, Ashanti, 27
determinism, genetic, 47–48
developing world, 224–45. *See also* indigenous people
  access to medications, 237–38
  health research for, 242
  pharmaceutical industry and, 239–42
  regenerative medicine and, 242–43
*Diamond, Commissioner of Patents and Trademarks v Chakrabarty,* 201–2
Dib, Nabil, 111–14
Diethrich, Edward, 113–15
disabilities, 48–50, 57–58
discrimination, genetic, 57, 73, 83
diseases, genetic, 7, 31–58, 75, 83. *See also specific diseases*
  complex, 33–34, 46, 62
  mutations as cause of, 33–34
  single-gene, 34, 46
DNA (deoxyribonucleic acid), 7, 32–33, 187. *See also* chromosomes; telomeres
  as commodity, 168, 216–17
  in Huntington's disease, 34
  of indigenous people, 224–27
  molecules of, 32–33
  and proteins, 33, 168–69
  recombinant, 167, 168–69, 170
  replication of, 44, 228
  three-base pairs in, 45
Doe, Tanis, 49
Dolly the sheep, 149, 190, 194
Dossetor, John, 145, 146
Down syndrome, 48–50, 107
drugs, 242. *See also specific drugs*
  abuse of, 92
  adverse reactions to, 84
  approval process of, 184
  "blockbuster," 185, 242
  clinical trials of, 184, 192, 193
  in developing world, 237–38
  development of, 173, 193
  generic, 238–41
  importation of, 238, 240–41
  licensing of, 238–40, 241
  patents on, 185, 192, 237–38
  pricing of, 237–38
Dugdale, Richard, 39

Dunston, Georgia, 226
During, Matthew, 8, 12

ears, 125–26
EGeen, 73
Ehrlich, Paul, 105
E.I. DuPont, 205–6
Ekstein, Joseph, 52
Eli Lilly & Co., 170, 210
embryos. *See also* abortion; stem cells, embryonic
  religious views on, 143–44, 158
  and research use, 107–8, 148, 159–65, 198
  rights of, 49–50
employers, 53–55, 57–58
Enfuvirtide (pentafuside), 174
engineering
  biomedical, 120
  genetic, 168, 199–204
  tissue, 116–42
enzymes. *See specific enzymes*
poetin alpha, 170
*Equal Employment Opportunity Commission (EEOC) v Burlington Northern Santa Fe Railroad Co.,* 57–58
Erickson, Robert, 20
erythropoietin (EPO), 92
Estonia, 70–73
ETC Group (Action Group on Erosion, Technology, and Concentration), 226–27, 230, 231–32
ethics, 144–46
  and clinical trials, 22–23
  culture and, 146–47
  and embryo use, 107–8, 198
  and patents, 205, 214–15
  research and, 25–26
  and society, 154–65
eugenics, 39–42
Europe
  *Convention on Human Rights and Biomedicine,* 72
  Directive on the Legal Protection of Biotechnological Inventions, 198
  and genetic information use, 56
  patents in, 198, 206, 214–15, 228
European Patent Convention, 198
European Patent Office, 228
euthanasia, 41–42

Faden, Ruth, 162
FDA. *See* United States Food and Drug Administration
Feldman, Marc, 226
Feng, Junli, 187
Fox, Michael, 144, 145, 149
France, 152
Fraser, Claire, 78
French Canadians, 75
FRISK Software International, 63
Frost, Norman, 108, 109
*Funk Brothers Seed Co. v Kalo Inoculant Co.,* 200
Futcher, Bruce, 179

Gage, Sanford, 219
Galton, Francis, 39
Ganguly, Arupa, 214
Garnier, Jean-Pierre, 241
Gearhart, John, 106–8, 162, 163, 186, 188, 195

# Index

Gelsinger, Jesse, 16–21, 28, 29
Gelsinger, Paul, 16–17, 18, 19–20, 24, 28–29
genealogical records, 60–61, 63–64, 65–69
GeneChip, 86–87, 89, 91
gene mapping, 34–35, 43–45, 75
  samples for, 65, 210
General Electric, 196–97, 199–204
genes, 6–7, 31–32. *See also* BRCA genes; proteins
  aldehyde dehydrogenase (ALDH), 84
  bone morphogenetic protein 2 (BMP2), 70
  commodification of, 69–70
  databases of, 91
  information about, 56, 83, 224
  mutations of, 33–34, 35, 38, 62, 74–75, 209–10
  p53, 186
  patents on, 187, 188, 210–11, 216
  research work on, 51–52, 91, 205, 212
  variations in, 81–85
Genetech, 84–85, 168–70, 204
gene therapy, 7–10
  clinical trials of, 17–21, 22, 27, 28
  databases on, 24, 30
  deaths from, 23
  FDA and, 11, 23–24
  National Institutes of Health (NIH) and, 23–24
  for ornithine transcarbamylase deficiency syndrome (OTC), 17–21, 22, 28
  for Parkinson's disease, 11–16, 244
  as proprietary process, 23–24
  regulation of, 26, 29–30
Gene Therapy Information Network (GTIN), 24
genetic engineering, 168, 199–204
geneticization, 47–48
genetic markers, 37–39, 43–44, 64–65
  restriction fragment length polymorphisms (RFLP), 34–35, 37, 209
genetic profiling, 225
genetics
  education in, 83
  population, 59–75, 82
Genetics Institute, Inc., 218
genome, 35, 79–80. *See also* Human Genome Project (HGP)
  human, 82–83, 86–87
Genome Canada, 83, 89
Genome Quebec, 83
GenomEUtwin, 75
genotypes, 81–85, 102
Genovo Inc., 28–29
Germany, 41–42
Geron Corp., 163, 165, 179–82, 186–95
  investors in, 186–87, 193
  products, 191–93
Gilbert, Wally, 210
Global Forum for Health Research, 242
Golde, David W., 217–22
government
  and academic-industry collaborations, 24–25
  inaction, 155–56
  policy development, 154–65
  research funding by, 162–63, 181
Gower, Jim, 169
Greely, Henry, 65–66
Greenberg, Herb, 172
Greenspan, Alan, 25
Greider, Carol, 179
Griffith, Fred, 32
Griffith-Cima, Linda, 126
GRN163, 191–92
Guaymi Indians, 230
"Guidelines for Good Clinical Practice" (International Conference on Harmonisation [ICH]), 21–22, 23
Gulcher, Jeffrey, 63, 64
Gusella, James, 36–38

haplotypes, 82–83
Harley, Cal, 179
Harry, Debra, 226, 227, 233
Harvard mouse. *See* OncoMouse
Harvard Stem Cell Institute, 164
Haseltine, William, 79, 90
Hatfield, Mark, 205
Hayflick, Leonard, 176–77, 178, 188
Health Canada, 23, 26, 184
health care, 54, 55, 66–67, 213
health insurance, 55
Health Sector Database (HSD), 65–69
heart, 110–15, 130–31
heart disease, 46, 110, 114
Heffron, Howard, 164
Herceptin, 84–85
Hereditary Disease Foundation, 34
hereditary neuropathy with liability to pressure palsies (HNPP), 58
heredity, 31–32, 36, 54
herpes simplex, 7–8, 9
HER2 protein, 84–85
Hitler, Adolf, 41–42
HIV, 8, 239–41, 242. *See also* AIDS
Hoffman, D.I., 159
Hoffmann-La Roche, 68, 70
Hospital for Sick Children (Toronto), 42, 43–45
human telomerase reverse transcriptase (hTERT), 188
Hubbard, Ruth, 206
Hudson, Thomas, 83
Human Genome Diversity Project (HGDP), 224–27
Human Genome Organization, 224
Human Genome Project (HGP), 76–81, 187–88
Human Genome Sciences (HGS), 78, 79, 216
Huntington's disease, 31, 34–39, 46, 50
Hyseq, 216

Iceland, 59–61, 62–70
  Health Sector Database Act (1998), 67, 69
  multiple sclerosis in, 62–63, 64–65
*The Immortal Cell* (West), 175
immune interferon 2, 217
Imperial Cancer Research Fund (London), 38
Incyte Genomics, 216
India, 51
indigenous people, 224–27, 228–30, 232–33. *See also* developing world
individualism, 147
Inglehart, Ronald, 154
initial public offerings (IPOs), 172, 182, 183–84, 204
Institut Curie, 212, 214–15
Institute for Cancer Research, 211

# Index


Institute for Cancer/Stem Cell Biology and Medicine, 164–65
The Institute for Genomic Research (TIGR), 78, 79
Institute for Populations and Genetics (IPEG), 74
insulin, 168–69, 170
insurers, 53, 55–57
intellectual property, 25, 163–64, 227
  rights to (IPRs), 235–41
interleukin-2, 217
International Conference on Harmonisation (ICH), 21, 23
International HapMap Project, 82–83
in vitro fertilization (IVF), 151–52, 160
IPOs (initial public offerings), 172, 182, 183–84, 204
Israel, 153

Jaenisch, Rudolf, 195
Japan, 25
Jefferson, Thomas, 197

Kaplitt, Martin, 5–6, 12–13
Kaplitt, Michael, 5–6, 7–10, 11–15
Karp, Jeffrey, 137–40
Kerem, Batsheva, 45
Kidd, Ken, 224
kidney failure, 92
King, Mary-Claire, 208–11, 224
Klein, Nathan, 3–5, 11–12, 13–15
Kleiner, Perkins, Caufield & Byers, 167, 169, 181
Kloiber, Otmar, 215
Knoppers, Bartha Maria, 72, 75
Krimsky, Sheldon, 206

Lambert, Janet, 184
Lane, Neal, 80
Langer, Robert S., 127, 128, 129
law
  and patents, 197–203
  and reproductive technology, 157–58
  and research, 25
lawsuits, 29, 49–50. *See also specific cases*
Leder, Philip, 205
leukemia, 217
Lewis, Stephen, 241
life insurance, 56
Lindpaintner, Klaus, 67–68
Lingner, Joachim, 187
Lippman, Abby, 47
Lipset, Seymour, 147
Living Implants From Engineering (LIFE), 130–31
Lou Gehrig's disease (amyotrophic lateral sclerosis, or ALS), 62
Love, James, 239
Lund, Dennis P., 129
Luo, Ying, 124–25
lymphoma, 174

MacDonald, Marcy, 38
malaria, 53, 242, 243
MaLossi, Leo I., 199–200
Mannvernd group, 69–70
Markey, Howard, 201
matREGEN, 117
McClung, Nellie, 41
McCormack, Sean, 129–30
McCulloch, Ernest, 94, 97–99, 100–102, 103–4, 106
McGill University, 210
Medicaid, 55
Medical Research Council (Canada), 26
Medical Research Council (U.K.), 73
Medicare (U.S.), 55
medications. *See* drugs
Melton, Doug, 164
Mendel, Gregor, 31–32
meningitis, 239
Merck & Co., 185, 233–34
mice (laboratory), 125–26, 162, 205–8
Millennium, 216
Miller, Loren, 228–29
Milstein, Alan, 21, 22, 23, 28, 29
Mooney, Pat, 230
Moore, John, 217–22
*Moore v Regents of the University of California,* 220–22
morality. *See* ethics
Morin, Gregg, 178–79
Morin, Yves, 156, 159
Mount Sinai Hospital (Toronto), 244
multiple sclerosis (MS), 62–63, 64–65, 93
muscle, 111, 130–31
muscular dystrophy, 93
mutations (genetic), 33–34, 35, 38, 62, 74–75, 209–10
myeloma, 191–92
myoblasts, 110–13
Myriad Genetics, 210–16

Nagy, Andras, 244
Nakamura, Toru, 188
National Bioethics Advisory Commission (U.S.), 153
National Commission for the Protection of Human Subjects of Biomedical and Behavioral Research (U.S.), 160
National Health Service Multi-Centre Research Ethics Committee (MREC), 74
National Institutes of Health (NIH)
  and gene therapy, 23–24
  and Gene Therapy Information Network (GTIN), 24
  and genetic information, 83, 224
  and Human Embryo Research panel, 160
  and Human Genome Project, 80
  and National Human Genome Research Institute (NHGRI), 77, 78, 82
  and Recombinant DNA Advisory Committee (RAC), 20–21, 23, 24, 30
  reporting requirements of, 23–24, 30
  research spending by, 164, 210
National Science Foundation (U.S.), 224
Natural Sciences and Engineering Research Council (Canada), 26
Nazis, 41–42
neem plant, 228
Negrette, Americo, 35–36
nerve growth factor (NGF), 244
nerve regeneration, 116–20, 122–25
nervous system, 118, 119
Neurologix, 12–13, 15
New York Weill Cornell Medical Center, 3
Nicaragua, 243
Noguchi, Philip, 24

# Index

Novartis, 227. *See also* Sandoz Pharmaceutical Group
Nuffield Council on Bioethics, 54, 216

O'Brien, Michelle and Paul, 48
Okarma, Thomas, 165, 190–91, 192, 193–94
Oliver, Richard W., 172
OncoMouse, 205–8
Ontario Cancer Institute, 97, 99–100
Ontario Health Insurance Plan (OHIP), 213–14
Organization for Economic Co-operation and Development (OECD), 47
organs
  regeneration of, 128–29
  transplants of, 115, 127–28, 158
ornithine transcarbamylase deficiency syndrome (OTC), 16, 17
  gene therapy trials for, 17–21, 22, 28
Osborn, Frederick, 42
osteoblasts, 134–35, 140
osteoclasts, 135
Osteofoam, 137
osteogenesis, 135
Ottawa General Hospital, 86
ovarian cancer, 209–10

Pálsson, Gísli, 69
*Parker v Flook,* 201
Parkinson's disease, 3–4, 9–16
  treatments for, 10–16, 244
patents, 196–223
  biotech industry and, 163–64, 193, 198, 212–15
  contravention of, 213–14, 215
  on drugs, 185, 192, 237–38
  ethical issues and, 205, 214–15
  on genes, 187, 188, 210–11, 216
  genetic engineering and, 199–204
  legislation of, 197–203, 204–7
  on living organisms, 205, 214–15
  monopolies created by, 212–15
  on plant use, 228–29, 232–33
  process claim of, 199, 206
  product claim of, 199–204
  for treatments, 185, 192
PAX7, 92–93
Pearson, Karl, 40–41
Perkin-Elmer Corp., 79, 80
Peterson, Rein, 180
Pfizer, 239
pharmaceutical industry, 184–86, 239–42
pharmacogenetics, 84, 173
phenotypes, 102
phenylketonuria (PKU), 47
Philippines, 230–31
plants, 228–29, 232–33, 234
plasmids, 167, 196–97
plasticity, 93–97
Poland's syndrome, 129
polyglycolic acid (PGA), 134
polylactic acid (PLA), 134, 138, 139
population genetics, 59–75, 82
preimplantation genetic diagnosis (PGD), 48–49, 51
Prendergast, Monsignor, 156
President's Council on Bioethics, 151–52, 158, 161, 163, 164, 165
Princess Margaret Hospital. *See* Ontario Cancer Institute
PRISM 3700, 79
privacy, 54–55, 56–57, 73, 83
prostate cancer, 192
proteins, 102, 123. *See also* genes
  bone morphogenetic (BMP), 134
  and cancer, 209
  on cell surfaces, 105–6
  DNA and, 33, 168–69, 170
  and drug reactions, 84
  HER2, 84–85
  and multiple sclerosis, 62
  Oct-4, 88
  p123, 187
Public Population Project in Genomics (P3G), 75

Quan, Shirley, 217–18
Quarles, Miller, 180–81
Quebec, 74–75, 83
Quigg, Donald J., 204–5

Rabinow, Paul, 69
radiation, 44, 100, 101, 102–3, 105
randomness, 103–4
Raper, Steven, 18, 19, 20, 21
Reddi, Hari, 134
regenerative medicine, 2, 90, 92–93, 105–6, 108–15, 116–42
  body's role in, 133–35
  for bone, 136–37
  developing world and, 242–43
  goal of, 120
  for organs, 128–31
regulation
  of cloning, 149–53, 155–59
  of gene therapy, 23, 26, 29–30
  of genetic testing, 47
  of research, 25, 26, 160–62
"Regulation of Gene Therapy: The Canadian Approach" (Ridgway), 23
religion
  and cloning, 153
  and embryo use, 143–44, 158
  and medical treatment, 146
  and reproductive technology, 156
  science and, 153–54
renal failure, 92
research
  in Canada, 24
  conflict of interest in, 25–26
  and developing world, 242
  funding for, 162–63, 181, 210
  in Japan, 25
  mice used in, 125–26, 162, 205–8
  by private companies, 162–63
  regulation of, 25, 26, 160–62
  results of, 91–93
  in United States, 24–25
  at universities, 24–26
research ethics boards (REBs), 25–26
Réseau de Médecine Génétique Appliquée (RMGA), 74
restriction fragment length polymorphism (RFLP) markers, 34–35, 37, 209
Retrovir, 238

# Index

retroviruses, 8
reverse transcriptase, 187
RiceTec, 232
Rich, Giles, 201
Ridgway, Anthony, 23, 26
Ridley, Matt, 34
rights, 147. *See also* abortion; ethics
  of embryo, 49–50
  of fetus, 51
  human, 155, 216–17
  of indigenous people, 227
  intellectual property (IPRs), 235–41
  privacy, 54–55, 83
  property, 216–22, 227
  of tissue donors, 216–22
Rituxan, 174
RNA (ribonucleic acid), 33, 187
Robbins-Roth, Cynthia, 169, 171
Robinson, Jeffrey, 239
Rockefeller Institute, 32
*Roe v Wade,* 160
Rommens, Johanna, 45
Roosevelt, Theodore, 40
Roslin Bio-Med, 194
Roslin Institute, 149, 190
Rothschild, David, 229
Royal Botanical Gardens (U.K.), 232–33
Royal Commission on New Reproductive Technologies, 155
Rozmahel, Richard, 45
Rudnicki, Michael, 86–88, 89–90, 92–93
Rural Advancement Foundation International (RAFI), 226–27. *See also* ETC Group
Ryan, Bill, 180

Samulski, Jude, 8
Sandel, Michael, 151
Sandoz Pharmaceutical Group, 218, 227. *See also* Novartis
Sanger Centre, 211
scaffolds
  for bone growth, 133, 134, 136–42
  for cartilage growth, 126, 130
  for nerve regeneration, 123–25
  three-dimensional, 128–29, 130
Schrecker, Ted, 146–47, 155
Schultes, Richard, 233
screening. *See* testing, genetic
Sefton, Michael, 130, 131
sex selection, 51
Shay, Jerry, 179, 186, 195
Shiva, Vandana, 229, 234
Shoichet, Brian, 120, 121–22
Shoichet, Molly, 2, 116–18, 120–25, 133, 136, 137
Shulman, Joel, 180
sickle-cell anemia, 53, 83
Silver, Lee, 144
Singapore Bioethics Advisory Committee, 152
single-nucleotide polymorphisms (SNPs), 81–82, 84
Skolnick, Mark, 210
Skúlason, Fridrik, 63
Slater, Samuel, 197
Social Sciences and Humanities Research Council, 26
society, 147
  ethics and, 154–65
Solow, Robert M., 171
somatic cell nuclear transfer (SCNT), 152, 158–59. *See also* cloning, therapeutic
Somerville, Margaret A., 49, 146–47, 155
South Africa, 240–41
South Korea, 149
Special Programme for Research and Training in Tropical Diseases (TDR), 242
spinal cord, 116–20, 122–25, 192–93, 244
Stanford University. *See* Institute for Cancer/Stem Cell Biology and Medicine
Stefánsson, Kári, 58, 61–70
Stefánsson, Stefan, 61–62
Stem Base, 91
Stem Cell Genomics Project, 89
Stem Cell Network, 89, 93–94, 95
stem cells, 86–115. *See also* cell lines; stem cells, embryonic
  adult, 89, 93–97
  expression profiling, 89–90, 91
  hematopoietic, 94, 106
  identification of, 101–2
  research on, 155–59, 181–82
  retinal, 244
  transplantation of, 105–6, 110–13, 141–42, 162
  as treatment, 91–92
stem cells, embryonic, 88–89, 143, 148
  isolating, 106, 107, 108
  views on using, 143–44, 145, 151–52
sterilization, forced, 40, 41
Stewart, Timothy, 205
stochastic theory, 103–4
Stolberg, Sheryl Gay, 7
Stoppa-Lyonnet, Dominique, 212, 215
Stratton, Michael, 211
stroke, 70
Strumpf, Robert, 115
Supreme Court of California, 220–21
Supreme Court of Canada, 51, 55, 207–8
Sustainable Science Institute, 243
Swanson, Bob, 167, 168–70

Tanzania, 243
TAQ polymerase, 228
Tay-Sachs disease, 46, 52
technology. *See also* biotech industry
  fear of misuse, 147–49
  reproductive, 155–59
telomerase, 109, 179–80, 181–82, 186–88, 190–92
telomeres, 177–79, 194. *See also* chromosomes; DNA
testing, genetic, 45–58
  and autonomy issues, 50, 51
  for breast cancer, 211–15
  consent for, 57–58
  costs of, 211–12, 213
  counselling for, 57, 211
  employers and, 53–55
  insurers and, 53, 55–57
  market for, 173, 211
  postnatal, 47, 51
  predictive, 46
  prenatal, 46–47, 48–52
  regulation of, 47
Thailand, 239
Third World. *See* developing world

# Index


Thomson, James, 108–9, 163, 181–82, 186, 188
Thorgeirsson, Thorgeir, 64–65
Thorsteinsdottir, Halla, 66–67
Till, James, 94, 98–100, 101–4
tissue engineering, 116–42
  for bone regeneration, 136–37
  for organ regeneration, 128–31
tracers (radioactive), 44
"Trade-Related Aspects of Intellectual Property Rights" (TRIPS), 236–41
transplants
  of cells, 105–6, 110–13, 141–42, 162
  of organs, 115, 127–28, 158
Triflucan, 239
Tropical Institute (Switzerland), 243
Tsui, Lap-Chee, 42–45

UCLA, 217–22
UK Biobank, 73–74
UNESCO, 56, 69, 72
United Kingdom, 73–74, 84, 152, 206, 238
United Nations, 216–17, 242
United States. *See also* United States: legislation
  abortion in, 160
  biotech industry in, 174
  breast cancer in, 53
  cloning regulation in, 150–52
  Department of Agriculture, 228
  Department of Energy (DOE), 77, 224
  Department of Health, Education and Welfare (DHEW), 160
  drug reactions in, 84
  economic sanctions imposed by, 240–41
  embryo research in, 159–65
  eugenics in, 39–41, 42
  and generic drugs, 239–41
  and genetic information use, 56
  health care in, 55
  heart disease in, 114
  Patent and Trademark Office (PTO), 199–200, 201, 203, 204–6
  patent legislation in, 197–203, 204–5
  research in, 24–25, 25, 160–62
  spinal cord injury in, 119–20
United States: legislation
  Americans with Disabilities Act (1990), 57–58
  Bayh-Dole Act (1980), 24
  Dickey Amendment (1995), 161–62
  Health Insurance Portability and Accountability Act (1996), 56–57
  Human Cloning Prohibition Act (2003), 151
  Patent Act (1793), 197
  Plant Patent Act (1995), 200
  Trade Act (1974, 1988), 240
United States Court of Custom and Patent Appeals (CCPA), 200, 201
United States Food and Drug Administration (FDA)
  drug approvals by, 184
  and gene therapy, 11, 23–24
  and Gene Therapy Information Network (GTIN), 24
  and ornithene transcarbamylase deficiency syndrome (OTC) trials, 20–21, 28–29
  reporting requirements of, 23–24
United States Supreme Court, 197–98, 200, 201–3
universities, 24–26. *See also specific universities*
University of California, 244. *See also Moore v Regents of the University of California*
University of Colorado, 187
University of Pennsylvania, 17–21, 22, 27–30, 214
University of Toronto, 116–18
University of Utah, 210
University of Vermont, 22
University of Wisconsin–Madison, 108, 163, 181, 228
Upton, Joseph, 129
Urist, Marshall R., 134
utilitarianism, 145–46

Vacanti, Charles, 125–27, 129–30
Vacanti, Joseph "Jay," 127–31
van der Kooy, Derek, 244
vectors, 7, 8–9, 12, 20, 26
Venezuela, 35–36, 38–39
Venter, Craig, 76, 77–81
vertebrae, 118, 119
viruses, 7–9, 12, 20. *See also specific viruses*

Wagner, Allen B., 218–19
Watson, James, 32–33, 76, 77, 85
Webster, Fiona, 213
Weissman, Irving, 104–6, 164–65
Wellcome Trust, 73
West, Michael, 175–76, 177–82, 186–87, 188–90, 194–95
Wexler, Alice, 39
Wexler, Leonore, 31
Wexler, Milton, 34
Wexler, Nancy, 31, 34, 35, 36–39
White, Ray, 44
White, Tony, 79–81
Wickelgren, Ingrid, 78–79
Williamson, Robert, 44, 49
Wilmut, Ian, 149
Wilson, Jack, 199
Wilson, James, 20, 21, 27–30
Winickoff, David E., 74
Winickoff, Richard N., 74
Wnt, 93
World Health Assembly, 241
World Health Organization (WHO), 57, 66
World Medical Association, 25
World Trade Organization (WTO), 236–41
Worton, Ron, 87, 156
W.R. Grace, 228
Wright, Woody, 178, 179
Wyngaarden, James, 77

Zevalin, 174
Zon, Leonard, 164